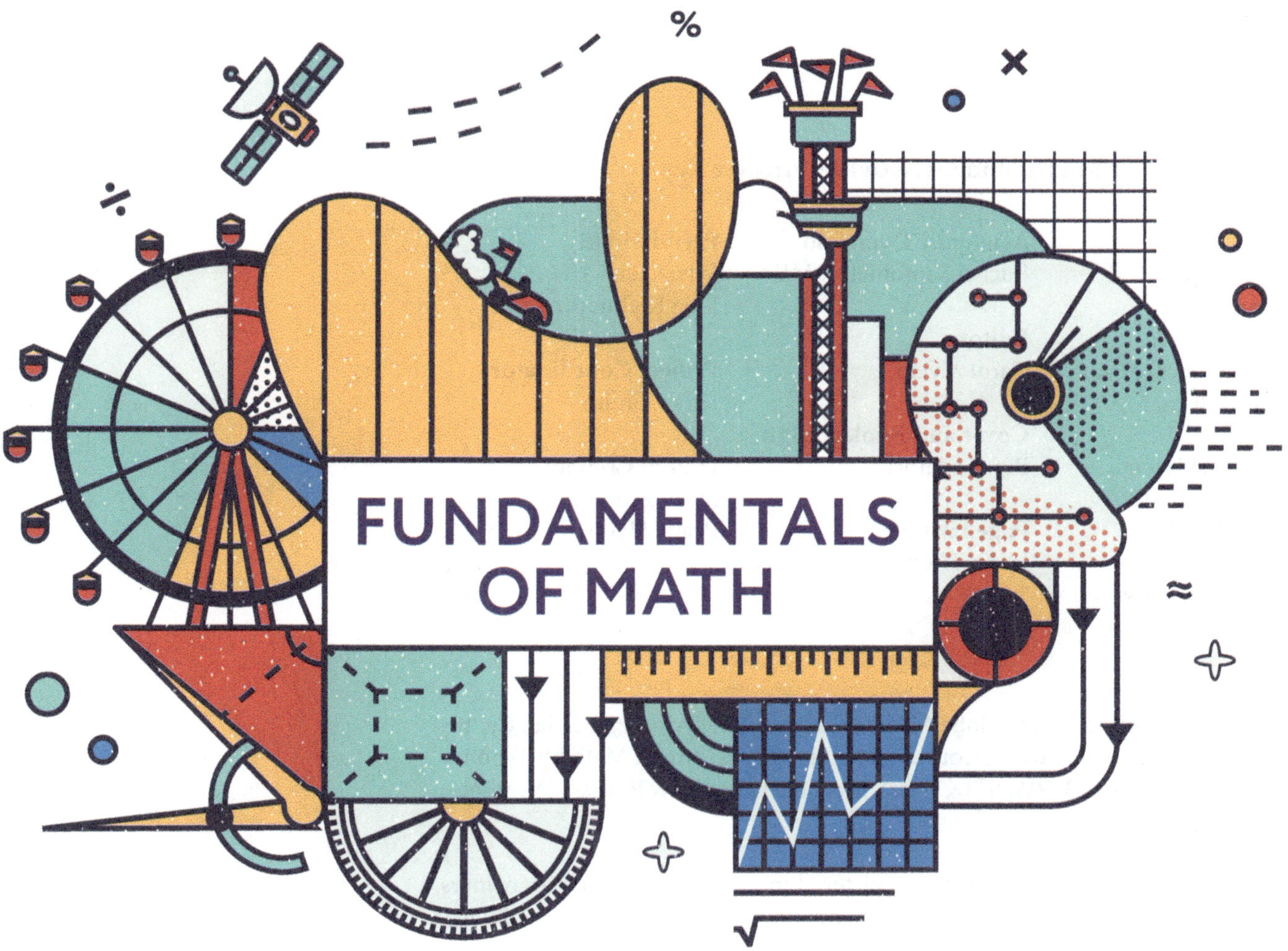

FUNDAMENTALS OF MATH

Third Edition

bju press®

Greenville, South Carolina

FUNDAMENTALS OF MATH Student Activities Answer Key
Third Edition

Writers
Tamera Knisely, MEd
Ben Adams, MEd
Timothy King
Donna Lawrence, MEd
Matthew Rollins, MA

Biblical Worldview
Tyler Trometer, MDiv
Bryan Smith, PhD

Academic Integrity
Jeff Heath, EdD

Academic Oversight
Rachel Santopietro, MEd

Editor
Carol Myers

Cover and Book Design
Brenna Short

Page Layout
Lydia Thompson

Permissions
Sylvia Gass
Lilia Kielmeyer

Project Coordinator
Kyla J. Smith

Digital Content Management
Peggy Hargis

The text for this book is set in Adobe Minion Pro, Adobe Myriad Pro, Arial, Calibri by Monotype, Circe by Paratype, Courier New, Helvetica, Minion Math by Typoma GmbH, Nimbus Sans by URW, STIX Fonts, Symbol, Times New Roman, Webdings, and Wingdings.

ISBN 978-1-62856-853-0

15 14 13 12 11 10 9 8 7 6 5 4 3 2 1

Introduction

Using This Book

This list of activities identifies the section being reviewed or the earliest section for which students have the appropriate background needed to do the activity. You may elect to use an activity at a later time, but it is not recommended that you use it sooner. Some of the activities will review several concepts. You may wish to do some of the activities over the course of several days.

Table of Contents

Adapting activities to your needs

You need not feel it necessary to cover every activity in this manual. Some may be used for remedial purposes for individual students; others you may wish to do as a class project. You can use parts of an activity without using the entire activity. Some of the activities review more than one section of the text—you may wish to do those in several sections. The purpose of the manual is to provide materials that will be helpful to you. You decide if, when, and how to use the activities.

Types of activities

Each chapter contains several of the possible types of activities. Included you will find enrichment, enhanced practice, extra practice, problem-solving, STEM, and collaborative activities.

1. Enrichment

Several chapters contain enrichment activities. You may wish to use them as class projects, assign them to students who desire to go beyond the required material, or use them as extra credit assignments.

2. Problem Solving

These activities contain word problems that relate to the chapter material.

3. Extra Practice (including Mixed Practice)

The extra practice activities are your basic drill sheets. Many topics covered in the textbook have additional review available through these activities. Several of these will review more than one section (mixed practice). The last activity in each chapter is a cumulative review of any material from that chapter or previous chapters. This could be used as a review activity or a pretest.

4. STEM

Detailed student instructions for each of the four STEM projects are provided. The extra teacher instructions for each STEM project are located at the back of the Teacher Edition.

Decimal Place Value
(Practice; use after Section 1.1)

Name ______________________________

A. 0.000196	**D.** 0.0196	**G.** 1.096	**J.** 1906	**M.** 109,060
B. 0.00196	**E.** 0.1096	**H.** 10.96	**K.** 10,960	**N.** 1,000,960
C. 0.01096	**F.** 0.196	**I.** 100.96	**L.** 19,600	**O.** 1,900,600

Match each word form with the standard form of the same number.

___H___ **1.** ten and ninety-six hundredths

___G___ **2.** one and ninety-six thousandths

___E___ **3.** one thousand ninety-six ten-thousandths

___M___ **4.** one hundred nine thousand, sixty

___C___ **5.** one thousand ninety-six hundred-thousandths

___O___ **6.** one million, nine hundred thousand, six hundred

___N___ **7.** one million, nine hundred sixty

___I___ **8.** one hundred and ninety-six hundredths

___J___ **9.** one thousand, nine hundred six

___K___ **10.** ten thousand, nine hundred sixty

___B___ **11.** one hundred ninety-six hundred-thousandths

___D___ **12.** one hundred ninety-six ten-thousandths

___F___ **13.** one hundred ninety-six thousandths

___L___ **14.** nineteen thousand, six hundred

___A___ **15.** one hundred ninety-six millionths

Comparing, Estimating, and Adding Whole Numbers

(Practice; reviews Sections 1.1–1.2)

Use the mileage chart below to answer exercises 1–22.

Name ______________________________

	Boston, MA	Dallas, TX	Denver, CO	Milwaukee, WI	New York, NY	Portland, OR	Washington, DC
Atlanta, GA	1037	672	1398	761	841	2601	608
Chicago, IL	994	917	996	87	802	2083	671
Indianapolis, IN	906	865	1058	268	713	2227	558
Jacksonville, FL	1155	990	1704	1067	959	2907	726
Los Angeles, CA	2979	1387	1059	2087	2786	959	2631
Omaha, NE	1412	644	537	493	1251	1654	1116
Philadelphia, PA	296	1452	1691	825	106	2821	143
San Francisco, CA	3095	1753	1235	2175	2934	636	2799

_______959 mi_______ 1. How many miles is it from Los Angeles to Portland?

_______1058 mi_______ 2. How many miles is it from Indianapolis to Denver?

_______2000 mi_______ 3. To the nearest thousand miles, how far is it from Dallas to San Francisco?

_______1200 mi_______ 4. To the nearest hundred miles, how far is it from Boston to Jacksonville?

_______760 mi_______ 5. To the nearest 10 miles, how far is it from Atlanta to Milwaukee?

_______3095 mi_______ 6. What is the greatest number of miles listed in the table?

_______87 mi_______ 7. What is the least number of miles listed?

_______Milwaukee_______ 8. Which city is closer to Atlanta—Milwaukee or New York?

_______Chicago_______ 9. Which city is farther from Washington—Chicago or Atlanta?

_______NY to San Francisco_______ 10. Is it farther from New York to San Francisco or from Portland to Philadelphia?

_______268; 558; 713; 906_______ 11. List the distances from Indianapolis to Boston, Milwaukee, New York, and Washington, DC, in order from least to greatest.

	Boston, MA	Dallas, TX	Denver, CO	Milwaukee, WI	New York, NY	Portland, OR	Washington, DC
Atlanta, GA	1037	672	1398	761	841	2601	608
Chicago, IL	994	917	996	87	802	2083	671
Indianapolis, IN	906	865	1058	268	713	2227	558
Jacksonville, FL	1155	990	1704	1067	959	2907	726
Los Angeles, CA	2979	1387	1059	2087	2786	959	2631
Omaha, NE	1412	644	537	493	1251	1654	1116
Philadelphia, PA	296	1452	1691	825	106	2821	143
San Francisco, CA	3095	1753	1235	2175	2934	636	2799

249 **12.** What is the smallest sum of two numbers in the row for Philadelphia?

New York
Washington, DC **13.** The answer for exercise 12 represents the distance between what two cities (by way of Philadelphia)?

3077 **14.** What is the second-largest sum of two numbers in the row for Chicago?

Boston & Portland **15.** The answer for exercise 14 represents the distance between what two cities (by way of Chicago)?

Omaha **16.** Passing through what city listed on the table will give you the shortest distance from Boston to Portland?

3066 mi **17.** How far is the shortest distance between Boston and Portland?

889 mi **18.** What is the shortest distance from Milwaukee to New York?

193; no **19.** Find the sum of the smallest two numbers in the table. Does the sum have meaning?

1874 mi **20.** It is 378 mi from Boston to Canada and 341 mi from Jacksonville to Miami. How far is it from Miami to Canada?

Los Angeles and San Francisco are on the Pacific Ocean. Boston, New York, and Jacksonville are on the Atlantic Ocean. (Washington, DC, is 116 mi inland.)

2377 mi **21.** What is the shortest distance from coast to coast?

LA to Jacksonville **22.** What cities give this distance?

Estimating Sums

Name ___________________________

Use estimation to solve these problems. Explain how you obtained each estimate.

1. John's mom sent him to the grocery store with a $10 bill to buy milk, bread, chips, eggs, and potatoes. He found milk for $2.49 per gal, bread for $0.89 per loaf, chips for $2.69 per bag, eggs for $0.79 per dz, and potatoes for $2.58 for 15 lb. Estimate the sum by rounding to the nearest dollar. Does John have enough money? Why are you certain?

   Milk is rounded to $2, bread to $1, chips to $3, eggs to $1, and potatoes to $3, totaling $10. Yes, he has enough money because when rounded to the nearest dollar, all the prices were rounded up except for the price of the milk, indicating that $10 is probably a high estimate.

2. Marcy has $10 to spend on snacks for a sleepover. She would like to buy a bag of chips for $2.09, 2 bottles of soda for $0.89 each, popcorn for $0.99, chip dip for $2.19, and brownie mix for $1.29. Estimate the sum by rounding to the nearest dime (tenth). Does Marcy have enough money left to buy a bag of candy for $1.49?

   The cost of chips is rounded to $2.10, soda to $1.80, popcorn to $1, dip to $2.20, and brownie mix to $1.30, totaling $8.40. Yes, she can purchase the bag of candy, since she has approximately $1.60 left.

3. Whitney has $20 to make a dress. She needs 4 yd of material at $3.98 per yard, a zipper for $1.49, and 2 yd of trim at $1.85 per yard. Estimate the cost by rounding to the nearest dollar. Does she have enough money to pay for the supplies?

   She will spend 4 times $4 ($16) on the material, $1 on the zipper, and 2 times $2 ($4) on the trim, totaling $21. No, she does not have enough for the supplies.

4. Alicia has $25 to use for her Christmas shopping. She is planning to buy an umbrella for her mom that costs $6.95. Her dad wants a Christmas CD that costs $5.69, and she has ordered a book for her sister that costs $6.38. Estimate her total expenditures to the nearest $0.10 and determine how much she has left to spend on her brother.

   She needs $7 for her mom, $5.70 for her dad, and $6.40 for her sister. This totals $19.10. She will have about $5.90 to spend on her brother.

Rounding Decimal Quotients

Jim and Tina are helping their youth pastor plan an outing to Pleasant Grove State Park. The park is located 107 mi from the church.

1. On the way to camp last month, the bus traveled 469 mi on 43 gal of gasoline. How many miles per gallon does the bus average (to the nearest tenth)?
   10.9 mi/gal

2. If the average speed for the bus is 50 mi/hr, how many hours will it take them to get to the park (to the nearest tenth)?   2.1 hr

3. Approximately how many gallons of gasoline will they need for the round trip (to the nearest tenth)?   19.6 gal

4. If gasoline costs $1.82 per gallon, how much money should they plan to pay for gas (to the nearest penny)? $35.67

5. A total of 25 young people are going on the activity. If they each paid $3.50, how much money will be left for food after they pay for the gasoline? $51.83

6. Tina went to the grocery store with the lunch menu and priced the following items. Find the total cost for the lunch items needed. Assume that there is no sales tax on groceries in their area. Try using the estimation method. Do they have enough money for this menu?
   - 6 packages of 8 hot dogs at $2.84 per package
   - 2 bottles of ketchup at $0.92
   - 1 bottle of mustard at $0.46
   - 1 jar of relish at $2.82
   - 4 packages of 12 hot dog buns at $2.92 each
   - 3 bags of chips at $2.98 each
   - 7 bottles of fruit punch at $0.99 each
   - 25 candy bars at $0.49 each
   - 1 bag of charcoal at $3.92

   total for groceries: $65.88; No, they are $14.05 short.

Calculating Camp Costs
(Practice; reviews Sections 1.2–1.4)

Name _______________________________

Victory Bible Camp is making plans for the upcoming camping season. The staff is planning for 320 campers per week for each of the 9 camping sessions. Each session lasts 5 days and costs $285 per camper.

Solve.

1. If each cabin holds 8 campers, how many cabins are needed each week to house the campers? 40 cabins

2. If the campers are split into 5 teams, how many campers will be on each team?
64 campers

3. How many cabins will it take to house each team? 8 cabins

4. The swimming pool can accommodate 56 swimmers at one time. To ensure that all cabins have a daily swimming time, how many different times must be scheduled? 6 swimming times

5. How much money will it cost each camper per day of camp? $57

6. How much money will the camp receive from all of the campers per week of camp? $91,200

7. How much money will the camp take in for the 9 sessions of the camping season?
$820,800

8. How many campers total are expected for the 9 camping sessions? 2880 campers

9. If the cafeteria can feed each camper for $15.50 per day, how much money should the camp director budget for food each day? $4960

10. How much money should the camp director budget for food for the campers per week? $24,800

Operations with Decimals

Name _______________________________________

Solve each problem. Be sure to label your answers.

1. Sofía and Aaron are working on a science project in which they have to keep track of the rainfall. During the first week of April it rained almost every day. It rained 0.3 in. on Monday, 1.2 in. on Tuesday, 0.09 in. on Thursday, 0.79 in. on Friday, and 0.15 in. on Saturday. What was the total rainfall for that week? 2.53 in.

2. Angela and her mother went shopping and purchased 4 packages of chicken. The packages weighed 2.96 lb, 3.2 lb, 2.875 lb, and 3.03 lb. How many pounds did they purchase? What is the answer rounded to the nearest tenth of a pound? 12.065 lb; 12.1 lb

3. The packages of chicken were priced $4.71, $5.09, $4.57, and $4.82. What was the total cost for the chicken? How much change did Angela's mother receive from $20? $19.19; $0.81

4. Angela was instructed to get exactly 4 lb of cheese. She picked up three packages weighing 1.37 lb, 1.18 lb, and 1.34 lb. Did she get enough cheese? If not, how many more pounds of cheese does she need to purchase? no; 0.11 lb

5. A 4-man relay team ran the mile relay in a time of 232 sec. A second team completed the first 3 legs of the race in 54.7 sec, 57.2 sec, and 61.3 sec. What must the fourth runner's time be in order for his team to beat the first team? less than 58.8 sec

Item	Price
sport coat	$79.95
dress shirt	$21.99
tie	$18.00
belt	$12.50
dress slacks	$37.75

6. Steve bought a new sport coat and dress pants. What was the total cost? If he put $25 down and charged the rest on a credit card for 3 months, what would his monthly payment be (not including interest on the borrowed money)? Cost: $117.70; Payment: $30.90/month

7. Miguel bought a dress shirt and tie. What was his total cost? He gave the clerk a $50 bill. What was the amount of change he received? Cost: $39.99; Change: $10.01

8. Jaquan wanted to buy a whole outfit. What would it cost him? He had only $110 in cash and decided to put the rest on his credit card. How much did he need to charge? Cost: $170.19; Charged: $60.19

9. Liam wants a pair of dress slacks, a belt, and a dress shirt. He has a $5 bill, two $10 bills and a $20 bill. How much money does he have? How much more will he need? Has: $45; Needs: $27.24

10. An ad in the paper has a coupon that gives $2 off each item purchased. If Liam uses the coupon, how much will his cost be for the items he wants? $66.24

Exponents

Name ________________________________

Write each in exponential form.

_____5^2_____ **1.** 5×5

_____8^5_____ **2.** $8 \times 8 \times 8 \times 8 \times 8$

_____13^3_____ **3.** $13 \times 13 \times 13$

_____3^4_____ **4.** $3 \times 3 \times 3 \times 3$

Write each in standard form.

_____1_____ **5.** 1^6

_____100_____ **6.** 10^2

_____729_____ **7.** 9^3

_____16_____ **8.** 2^4

_____243_____ **9.** 3^5

_____1728_____ **10.** 12^3

_____16_____ **11.** 4 squared

_____343_____ **12.** 7 cubed

_____169_____ **13.** 13 squared

_____1000_____ **14.** 10 cubed

Find the products.

_____50_____ **15.** $2 \cdot 5^2$

_____375_____ **16.** $3 \cdot 5^3$

_____70,000_____ **17.** $7 \cdot 10^4$

_____64_____ **18.** 4×2^4

_____576_____ **19.** $3^2 \times 4^3$

_____15_____ **20.** $15^1 \times 1^{15}$

Approximating the Root Cause

(Enrichment; use after Section 1.6)

In the chapter, you learned that you could approximate the value of a square root by determining which two perfect squares the value under the radical falls between.

1. For example, to calculate $\sqrt{14}$, first determine between which two perfect squares 14 falls. 9 and 16

2. So, $\sqrt{14}$ falls between what two numbers? Which one is it most likely closest to? Why?
   3 and 4; closer to 4 since 14 is closer to 16 than 9

3. To approximate square roots to the nearest tenth, you would normally use a calculator. However, we want to explore how students found a square root before the invention of calculators. Calculate the following:

   $3.9 \times 3.9 = \underline{\quad 15.21 \quad}$ $3.7 \times 3.7 = \underline{\quad 13.69 \quad}$

   $3.8 \times 3.8 = \underline{\quad 14.44 \quad}$ $3.6 \times 3.6 = \underline{\quad 12.96 \quad}$

4. Which of those does 14 fall between? Is it closer to one or the other, or is it in the middle? 3.7 and 3.8; in the middle

5. To find a closer estimate, include one more digit behind the decimal. Find each product using your calculator:

   $3.73 \times 3.73 = \underline{\quad 13.9129 \quad}$ $3.74 \times 3.74 = \underline{\quad 13.9876 \quad}$

   $3.75 \times 3.75 = \underline{\quad 14.0625 \quad}$ $3.76 \times 3.76 = \underline{\quad 14.1376 \quad}$

6. Between which two factors from exercise 5 is 14 located? Is it closer to one number or the other—or is it in the middle? 3.74 and 3.75; closer to 3.74

7. Now try using the square root button on your calculator. Find the square root of 14, and round it to the nearest thousandth. 3.742

8. You could continue this process to find as many decimal places as you wish! Try again. This time find $\sqrt{54}$ to the nearest hundredth, and then check your answer with the square root button on your calculator. Show your work. 54 is between 49 and 64—but closer to 49.

   So, $\sqrt{54}$ is between 7 and 8—closer to 7. $7.3^2 = 53.29$ and $7.4^2 = 54.76$, so $\sqrt{54}$ is between 7.3 and 7.4, toward the middle. $7.34^2 = 53.8756$ and $7.35^2 = 54.0225$, so $\sqrt{54}$ is between 7.34 and 7.35—but closer to 7.35. Actual $\sqrt{54} = 7.348$.
   Keyword search: radical table math; slide rule

Order of Operations
(Practice; use after Section 1.7)

Name _______________________________________

Evaluate by following the order of operations.

___7___ **1.** $24 \div 8 + 4$

___16___ **2.** $13 + 5 - 2$

___25___ **3.** $7 + 6 \times 3$

___33___ **4.** $28 + 15 \div 3$

___8___ **5.** $36 \div 9 \times 2$

___5___ **6.** $20 - 5 \times 3$

___6___ **7.** $18 \div 6 + 3$

___12___ **8.** $4 \times 6 \div 2$

___7___ **9.** $5 + (7 - 3) \div 2$

___25___ **10.** $18 + 2 \times 6 - 15 \div 3$

___20___ **11.** $(12 - 4) \div 2 + 16$

___28___ **12.** $30 - 3 \times 4 + 2 \times 5$

___85___ **13.** $(5 \times 8) + 3 \times 15$

___19___ **14.** $(9 + 3) \times 3 \div 2 + 1$

___25___ **15.** $24 + 18 \div 3 - 5$

___7___ **16.** $56 \div 7 + 3 - 8 \div 2$

___10___ **17.** $(32 + 13) \div 15 + 7$

Practice with Sets
(Extra Practice; use after Section 2.1)

Name _______________________________

Choose the correct set description.

___B___ **1.** {1, 2, 3, 4, 5, 6}

        A. {natural numbers less than 6}
        B. {positive integers less than 7}
        C. {integers between 1 and 6}
        D. not given

Use the Venn diagram for exercises 2–4.

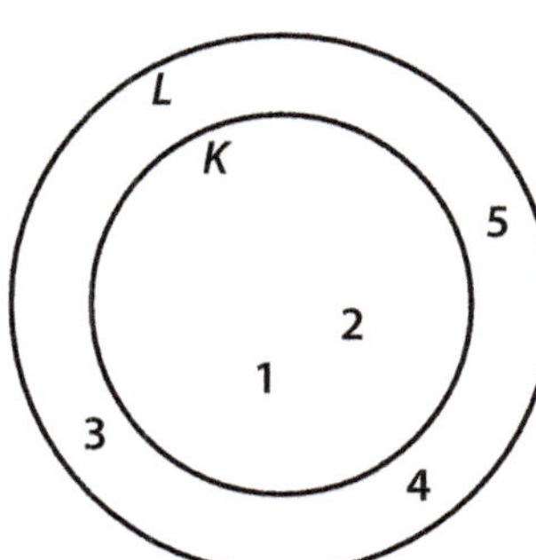

$K = \{1, 2\}$ **2.** List the members of set K.

$L = \{1, 2, 3, 4, 5\}$ **3.** List the members of set L.

$K \subseteq L$ **4.** Show the relationship of K and L using set notation.

Determine which choice is false for the given sets.

___D___ **5.** $A = \{2, 4, 8, 16, 32\}$
        $B = \{2, 4, 8, 16, 32 \ldots\}$

        A. $8 \notin A$
        B. $64 \in B$
        C. $64 \notin A$
        D. $8 \notin B$

___A___ **6.** $A = \{1, 2, 3\}$, $B = \{2, 3\}$, and
        $C = \{2, 3, 4, 5, 6\}$

        A. $A \subseteq B$
        B. $B \subseteq A$
        C. $B \subseteq C$
        D. $\varnothing \subseteq C$

Choose the correct answer.

___C___ **7.** Which set is infinite?

        A. {natural numbers less than 200}
        B. {integers between −5 and 5}
        C. {rational numbers between 1 and 2}
        D. All the given sets are finite.

___D___ **8.** Which of the following is not true?

        A. $3 \in \mathbb{Z}$
        B. $\frac{3}{4} \in \mathbb{R}$
        C. $0 \notin \mathbb{N}$
        D. $-2 \notin \mathbb{Q}$

  9. Which is not a subset of $\mathbb{Z}$?

      A. {−2, 3, 4, 26}
      B. {1, 2.75, 12, 300}
      C. {−5, −3, −2}
      D. {0}

  10. Determine whether the subset sequence is true or false. $\mathbb{N} \subseteq \mathbb{W} \subseteq \mathbb{Z} \subseteq \mathbb{Q} \subseteq \mathbb{R}$

Subsets

Name ______________________________

Example: List all the subsets for {a, b}.

Answer: { } or Ø, {a}, {b}, {a, b}

1. List the eight subsets for {a, b, c}. Ø, {a}, {b}, {c}, {a, b}, {a, c}, {b, c}, {a, b, c}

Officer Jones is assigned to control the traffic at 4 intersections. To make sure that his location is not predictable, his supervisor made a list using all the different combinations of the 4 intersections. On a given day, Officer Jones may visit 1, 2, 3, or all 4 of the intersections. (In other words, he might stay in the same place all day, or he might move to all 4 locations on the same day.)

Determine the possible locations by listing all the possible combinations. An "element" in this case is one of the intersections (1, 2, 3, or 4) where Officer Jones might be stationed on a given day.

2. subsets with 1 element: {1}, {2}, {3}, {4}

3. subsets with 2 elements: {1, 2}, {1, 3}, {1, 4}, {2, 3}, {2, 4}, {3, 4}

4. subsets with 3 elements: {1, 2, 3}, {1, 2, 4}, {1, 3, 4}, {2, 3, 4}

5. subsets with 4 elements: {1, 2, 3, 4}

6. How many different combinations of intersections are possible for Officer Jones? 15

Comparing Integers and Absolute Value

(Extra Practice; use after Section 2.2)

Write an integer that expresses the meaning of the given phrase.

−50 **1.** a $50 expense

800 **2.** $800 income

0 **3.** mean sea level

5 **4.** 5 m high

32 **5.** 32 degrees above 0

6194 **6.** 6194 m above sea level

−55 **7.** a 55-point drop in the stock market

−10 **8.** 10 sec to blastoff

Place the lists of integers in order from least to greatest.

9. $2, 0, -2, 3, -5, 1$ $-5, -2, 0, 1, 2, 3$

10. $23, 32, -16, 30, -18, -4$ $-18, -16, -4, 23, 30, 32$

11. $-8, -12, -1, -14, -9, -5, 0$ $-14, -12, -9, -8, -5, -1, 0$

12. $2^3, 4, 4^2, 4^3, 2, 2^5$ $2, 4, 2^3, 4^2, 2^5, 4^3$

13. $-4^3, -2^2, -2^4, -4^5, -2^5, -2$ $-4^5, -4^3, -2^5, -2^4, -2^2, -2$

Simplify the expressions containing absolute values, giving special attention to the order of operations.

5   **14.** $18 - |-13|$

3   **15.** $|23 - 14| \div |-3|$

12   **16.** $|-42| \cdot 2 \div |-7|$

12   **17.** $|16 - 2 \cdot 4| + |12 \div -3|$

185   **18.** $|45 - 2^3| \cdot |-5|$

26   **19.** $22 + |36 \div -9|$

0   **20.** $24 - 18 \div |-3| \cdot 4$

0   **21.** $|-64 \div 16 \cdot 2| - 4|82 - 7 \cdot 12|$

2   **22.** $\dfrac{|32 - 6|}{|-13|}$

−5   **23.** $\dfrac{|8 - 13| + 3|-5|}{-|4|}$

−23   **24.** $|-8 \cdot 2| - |-13| \cdot |-3|$

24   **25.** $21 + 5|-15| \div |-5^2|$

Adding and Subtracting Integers

(Extra Practice; use after Section 2.4)

Name _______________________________

Add.

_____1_____ **1.** $4 + (-3)$

_____−4_____ **2.** $-8 + 4$

_____2_____ **3.** $-7 + 9$

_____11_____ **4.** $-2 + 13$

_____−9_____ **5.** $-6 + (-3)$

_____−12_____ **6.** $-5 + (-7)$

_____−3_____ **7.** $9 + (-12)$

_____−5_____ **8.** $-1 + (-4)$

_____−2_____ **9.** $-14 + 12$

_____−3_____ **10.** $1 + (-4)$

_____2_____ **11.** $-15 + 17$

_____−12_____ **12.** $-4 + (-8)$

_____−21_____ **13.** $-9 + (-12)$

_____−9_____ **14.** $2 + (-11)$

Subtract.

_____−2_____ **15.** $6 - 8$

_____5_____ **16.** $0 - (-5)$

_____9_____ **17.** $4 - (-5)$

_____−15_____ **18.** $-9 - 6$

_____5_____ **19.** $-2 - (-7)$

_____−14_____ **20.** $-6 - 8$

_____−3_____ **21.** $-5 - (-2)$

_____12_____ **22.** $9 - (-3)$

_____16_____ **23.** $9 - (-7)$

_____−7_____ **24.** $5 - 12$

_____−5_____ **25.** $6 - 11$

_____0_____ **26.** $-3 - (-3)$

Add or subtract as indicated.

_____14_____ **27.** $-6 + 15 - 3 + 8$

_____−10_____ **28.** $-8 + (-9) - (-7)$

_____0_____ **29.** $5 - 11 - (-6)$

_____−21_____ **30.** $-17 + 5 - 8 + 3 + (-4)$

Write each problem as a sum or difference of integers and then solve.

1. In a single day, the temperature rose from −9°F to 10°F. Find the change in the temperatures. 19°F

2. From 6 PM to 10 PM the temperature dropped from 42°F to 18°F. What was the average temperature change per hour? −6°F

3. The rainfall for the month was 1.25 in. above the average of 0.9 in. How much rain fell during the month? 2.15 in.

4. A toy store reported a $3120 profit in April, a $1984 loss in May, and a profit of $2487 in June. Find the profit or loss for the quarter? $3623

5. Ben earned $519 last month. He charged $605 on his credit card. How much did he overcharge? −$86

6. Dave paid $728 in rent and $891 in other bills. He earned $2562 this month and has $1537 in savings. After paying his bills, the remainder is deposited into savings. How much is now in his savings account? $2480

7. A small business ended the year in debt for $19,820, but the next year they earned $169,104. After paying off the debt, they paid $72,948 for salaries and $54,803 for other costs. Find the end of the year balance for the company. $21,533

8. The Panthers football team began at the 5 yd line and gained 25 yd on the 1st play, lost 5 yd on the 2nd play, and gained 10 yd on the 3rd play. They gained 7 yd on the 4th play, lost 6 yd on the 5th play, and gained 12 yd on the 6th play. On which yard line would the Panthers be after the 6th play? 48 yd line

9. The Pax Romana (Roman peace) was the period of time during which Jesus had His earthly ministry and His disciples began to spread the good news to "the ends of the earth." The Pax Romana is dated from 30 BC to AD 180. How long did the Pax Romana last? 210 years

Name _______________________________________

Write the missing sign (+ or −) in the box.

1. $-8 + (\boxed{+}\,13) = 5$

2. $\boxed{+}\,4 + (-1) = 3$

3. $\boxed{-}\,7 + (-12) = -19$

4. $-14 - (\boxed{+}\,5) = -19$

5. $+15 + (\boxed{-}\,8) = 7$

6. $+17 - (\boxed{-}\,5) = 22$

7. $+7 - (\boxed{-}\,3) = 10$

8. $\boxed{-}\,4 \cdot 8 = -32$

9. $\boxed{-}\,6 - 9 = -15$

10. $-7 \cdot \boxed{-}\,3 = 21$

11. $\boxed{+}\,12 + (-4) = 8$

12. $-4 + (+2) + (\boxed{+}\,5) = 3$

13. $\boxed{-}\,9 \cdot (-6) = 54$

14. $-8 + (\boxed{-}\,5) + 4 = -9$

15. $+7 \cdot \boxed{-}\,3 = -21$

16. $+5 - (8) + 0 + (\boxed{-}\,10) = -13$

17. $-5 \cdot \boxed{+}\,8 = -40$

18. $2 \cdot (-3) \cdot (\boxed{+}\,4) = -24$

19. $18 \div \boxed{-}\,9 = -2$

20. $-2 \cdot \boxed{+}\,3 \cdot (-8) = 48$

21. $\boxed{-}\,63 \div (-7) = 9$

22. $\boxed{+}\,36 \div (-3) \div (-2) = 6$

23. $-21 \div \boxed{+}\,3 = -7$

24. $-6 \cdot \boxed{-}\,5 \div 3 = 10$

25. $+3 + (\boxed{-}\,7) = -4$

Amazing Integer Review
(Mixed Practice; use after Section 2.6)

Find your way through the maze by choosing the bridges with the correct answers.

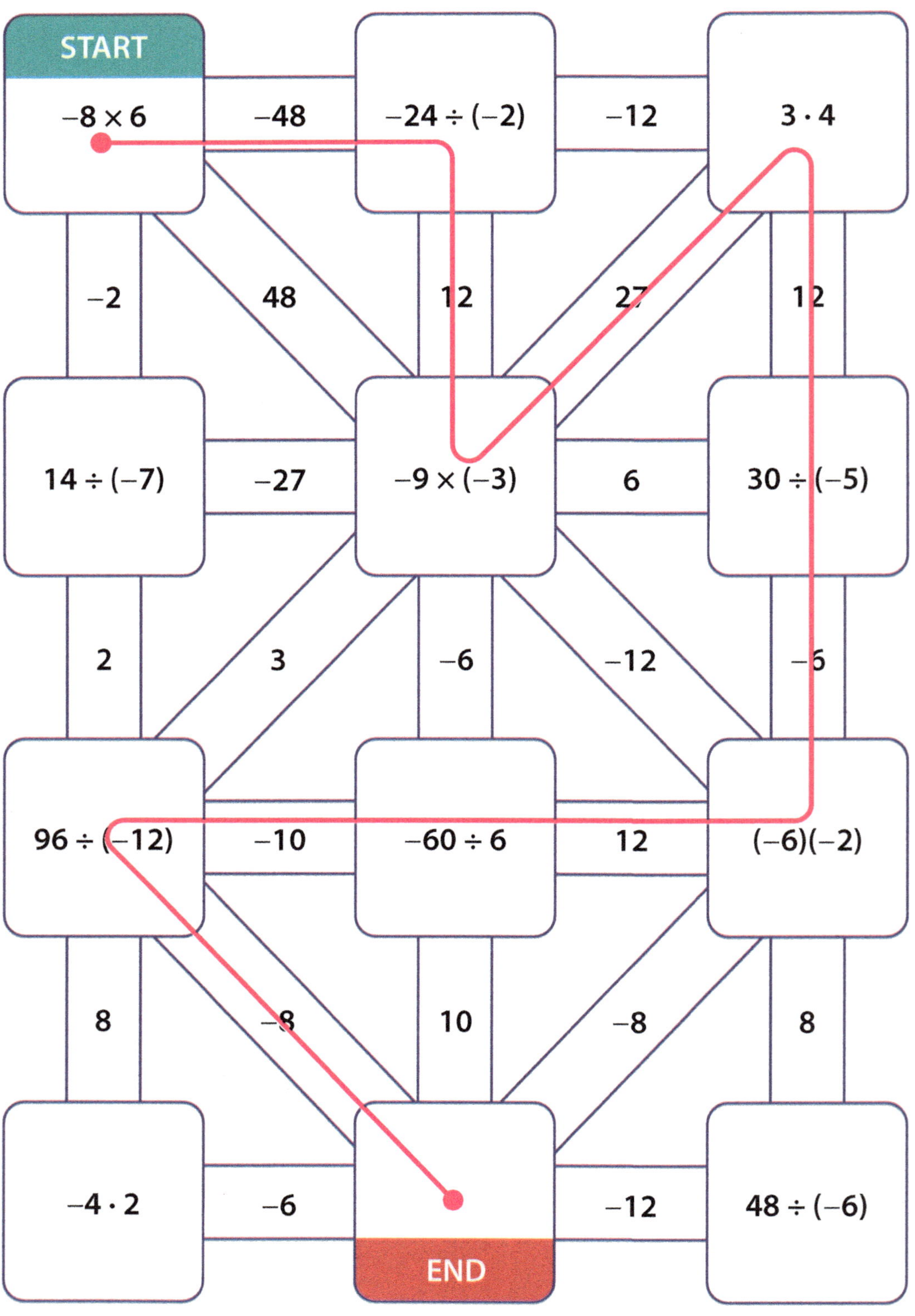

Fundamentals of Math

Name ______________________________

_____B_____ **1.** What is the standard form for three million, two hundred thousand, nineteen? **[1.1]**

    A. 3,020,019
    B. 3,200,019
    C. 3,002,019
    D. not given

_____C_____ **2.** Round 1,506,293 to the nearest ten thousand. **[1.1]**

    A. 1,500,000
    B. 1,506,000
    C. 1,510,000
    D. not given

_____C_____ **3.** Which numbers are in decreasing order? **[1.1]**

    A. 2403; 2430; 2340; 2304
    B. 2304; 2340; 2403; 2430
    C. 2430; 2403; 2340; 2304
    D. not given

_____A_____ **4.** Add $9375 + 6418 + 7206$. **[1.2]**

    A. 22,999
    B. 23,099
    C. 22,989
    D. not given

_____C_____ **5.** Perform the following operations:
$312.79 - 28.9432 - 7.93$. **[1.2]**

    A. 283.8532
    B. 276.9168
    C. 275.9168
    D. not given

_____ B   **6.** Multiply 724×385. **[1.3]**

      A. 277,740
      B. 278,740
      C. 278,640
      D. not given

_____ A   **7.** Divide $32{,}752 \div 89$. **[1.4]**

      A. 368
      B. 298
      C. 378
      D. not given

_____ D   **8.** Find $\sqrt{289}$. **[1.6]**

      A. 13
      B. 23
      C. 19
      D. not given

_____ C   **9.** Simplify $2.5 + 3.1 \times 5$. **[1.7]**

      A. 28
      B. 157.5
      C. 18
      D. not given

_____ A   **10.** Calculate $800 - (623 - 157)$. **[1.7]**

      A. 334
      B. 20
      C. 434
      D. not given

Algebraic Expressions
(Extra Practice; use after Section 3.1)

Name ________________________

Match each phrase with the appropriate algebraic expression.

A. $n - 6$	**E.** $\frac{3n}{2}$	**H.** $n + 6$
B. $\frac{n}{2} - 3$	**F.** $2n + 3$	**I.** $\frac{6}{n}$
C. $\frac{n}{6}$	**G.** $\frac{2n}{3}$	**J.** $3n + 2$
D. $6n$		

___H___ **1.** 6 more than n

___D___ **2.** 6 times a number

___F___ **3.** 3 more than twice a number

___A___ **4.** 6 less than n

___C___ **5.** a number divided by 6

___B___ **6.** 3 less than the quotient of n and 2

___J___ **7.** 2 more than 3 times n

___G___ **8.** twice a number divided by 3

___I___ **9.** 6 divided by a number

___E___ **10.** 3 times a number, divided by 2

Write an expression for each phrase.

___$n + 5$___ **11.** 5 more than a number n

___$\frac{q}{4}$, or $q \div 4$___ **12.** a number q divided by 4

___$3w - 7$___ **13.** a number w tripled, minus 7

___$8m - 9$___ **14.** 8 times a number m, decreased by 9

___$18 - 4r$___ **15.** 18 minus the product of 4 and a number r

___$4 - p$___ **16.** the difference of 4 and a number p

___$\frac{n}{6} + 5$___ **17.** 5 more than the quotient of a number n and 6

___$13s$___ **18.** the product of 13 and a number s

___$2(n + 4)$___ **19.** 2 times the sum of a number n and 4

___$\frac{3}{4}z$___ **20.** three-fourths of a number z

Using Algebraic Expressions
(Mixed Practice; use after Section 3.1)

Refer to the table below to write algebraic expressions for each of the following.

Fall Soccer Tournament				
Team	Eagles	Patriots	Bulldogs	Warriors
Number of Players	15	12	14	n

$n + 15$ **1.** the number of players on the Warriors and the Eagles

$12 - n$ **2.** the difference between the number of players on the Patriots and the number of players on the Warriors

$3n$ **3.** 3 times the number of Warriors players

$2n - 15$ **4.** twice the number of Warriors players less the number of Eagles players

$(15 + 14) + n$ **5.** the total number of players on both the Eagles and the Bulldogs, increased by the number of players on the Warriors

$3 \cdot 14 - n$ **6.** 3 times the number of Bulldogs players, decreased by the number of Warriors players

$3(n + 9)$ **7.** 9 more than the number of Warriors players, multiplied by 3

$\dfrac{4 \times 15}{n}$ **8.** 4 times the number of Eagles players, divided by the number of Warriors players

$\dfrac{12}{n} + 5$ **9.** the quotient of Patriots and Warriors players, increased by 5

$\dfrac{12 + 14 + n}{3}$ **10.** the sum of Patriots, Bulldogs, and Warriors players, divided by 3

Cooperative Learning: Soccer Field Borders

Name _______________________

(Extra Practice; use after Section 3.1; revisit after Sec. 3.5 to include the other lengths)

Write your answers for each section on a separate sheet of paper. *See note at the bottom of the page.

You are a business owner creating a soccer complex with fields of various sizes.

Big Question: How much marking chalk or paint is required to make the lines of each type of youth soccer field? At what cost? (Include only the 4 boundary lines and the halfway line.) ①

Usually, line-marking machines have a set width of about 4 inches ($\frac{1}{3}$ ft). We are measuring the amount of marking chalk or paint in linear feet only, not square feet. ②

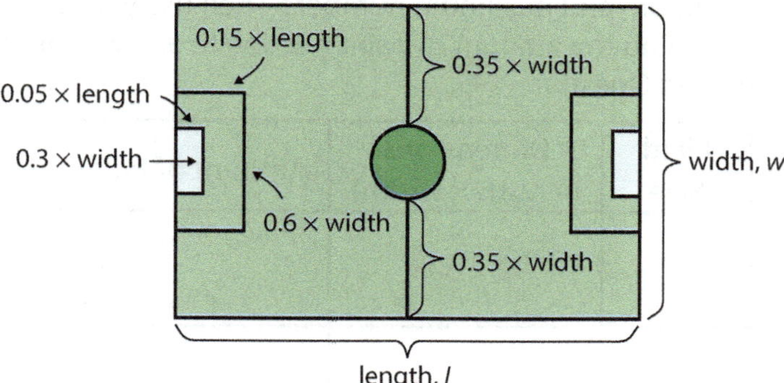

To complete this activity, you must:

- Write an expression modeling how many feet of paint are required for marking a field of any length and width in general. ③

- Use this expression to find how much paint is needed to mark each type of youth field (U6, U8, U10, U12).

- Determine the number of 20 oz cans of marking paint required to line each type of youth field. How much would each field cost to mark up?

BRAINSTORM

Answer the questions below *individually*, then discuss them when joining your group.

1. What is the question I am exploring? How much paint do I need to make the lines for a soccer field?

2. What is the expression that represents the amount of paint needed to mark up any field? Why? There are 3 lines measuring w yards. Then there are 2 full lengths l. So, we have $l + l + w + w + w$, or $2l + 3w$.

3. What facts do I need to know to answer all the parts of the big question? dimensions of each field type, amount of paint for each field, how much a container of marking paint covers

COLLABORATE

Share your "Brainstorm" answers with your partner(s) and discuss your interpretations of the problem. When your group comes to an agreement on what needs to be accomplished and what information needs to be researched, finish the following sentences.

1. The important quantities in this situation are . . .
   the dimensions of the field, the number of fields, how much a can of marking paint covers

2. It will be helpful to know . . .

3. We think the expression modeling this situation is . . . because . . .

4. Write down 3 questions you need to answer to completely answer the Big Question.

*Additional teacher notes for this activity, matched with the circled numbers, are provided in the back of the Teacher Edition.

RESEARCH & EVALUATE USING YOUR MODEL

Follow your teacher's instruction to find dimensions for each field size below (each group member can complete the work for 1–2 fields) and fill in the table.

(If you are given a range of sizes, be sure to find the average of the minimum and maximum numbers. Then use your general expression modeling the amount of paint needed to answer the rest of the parts from the Big Question.) ④

Field Size	Dimensions (length × width)	Amount of Paint Needed	Number of Spray Paint Cans Needed
U6	90 × 60 ft	2(90) + 3(60) = 360 ft	360 ÷ 200 = 1.8 cans 360 ÷ 300 = 1.2 cans
U8	90 × 60 ft	360 ft	1.8 cans if it covers 200 ft, 1.2 if it covers 300 ft
U10	180 × 120 ft	2(180) + 3(120) = 720 ft	720 ÷ 200 = 3.6 cans 720 ÷ 300 = 2.4 cans
U12	225 × 150 ft	2(225) + 3(150) = 900 ft	900 ÷ 200 = 4.5 cans 900 ÷ 300 = 3 cans

⑤

1. How much paint do you need if you have one of each field?

   360 + 360 + 720 + 900 = 2340 ft of paint (see table above)

2. How much paint do you need if you have three U8 fields and two U12 fields?
   3(360) + 2(900) = 2880 ft (see table above)

3. Calculate the cost of marking each type of field.

4. Why did you use the numbers you did? ⑥

EXTENSION

You own a 450 × 300 yd rectangular plot of land.

- Draw the space on your separate sheet of paper.
- With your partner, lay out as many youth fields as you wish to fit in the space, with any orientation. Draw and label them (U6, U8, etc.) in the space.
- Choose the sizes for each field and determine the spacing between them.

Answer the questions below.

1. How much paint is needed to cover all of these fields?

2. How many cans of paint are needed, and what is the cost?

3. As a businessperson looking to make money, why did you choose to lay out the space the way you did? (Include the number of fields, the cost, why you think this will make you money, whom this layout serves, and so on.)

Properties of Addition

(Extra Practice; use after Section 3.2)

Name _______________________________________

Use the Commutative Property of Addition to fill in the missing constant or variable.

1. $8 + 5 = 5 + \underline{\hspace{1em} 8 \hspace{1em}}$

2. $7 + 12 = \underline{\hspace{1em} 12 + 7 \hspace{1em}}$

3. $n + 23 = \underline{\hspace{1em} 23 + n \hspace{1em}}$

4. $x + y = \underline{\hspace{1em} y \hspace{1em}} + x$

Use the Associative Property of Addition to fill in the missing constant or variable.

5. $9 + (4 + 13) = (9 + 4) + \underline{\hspace{1em} 13 \hspace{1em}}$

6. $(6 + 7) + 3 = \underline{\hspace{1em} 6 \hspace{1em}} + (7 + 3)$

7. $5 + (x + 8) = (5 + \underline{\hspace{1em} x \hspace{1em}}) + 8$

8. $(r + s) + v = r + \underline{\hspace{1em} (s + v) \hspace{1em}}$

Use the Identity Property of Addition to fill in the missing number.

9. $796 + 0 = \underline{\hspace{1em} 796 \hspace{1em}}$

10. $y + \underline{\hspace{1em} 0 \hspace{1em}} = y$

Identify the property of addition illustrated. Use *A* for Associative, *C* for Commutative and *I* for Identity.

$\underline{\hspace{1em} A \hspace{1em}}$ **11.** $8 + (2 + 5) = (8 + 2) + 5$

$\underline{\hspace{1em} I \hspace{1em}}$ **12.** $0 + 175 = 175$

$\underline{\hspace{1em} C \hspace{1em}}$ **13.** $17 + 8 = 8 + 17$

$\underline{\hspace{1em} C \hspace{1em}}$ **14.** $d + 17 = 17 + d$

$\underline{\hspace{1em} I \hspace{1em}}$ **15.** $0 + m = m$

$\underline{\hspace{1em} A \hspace{1em}}$ **16.** $n + (3 + 5) = (n + 3) + 5$

$\underline{\hspace{1em} C \hspace{1em}}$ **17.** $r + (s + t) = r + (t + s)$

$\underline{\hspace{1em} A \hspace{1em}}$ **18.** $(p + q) + r = p + (q + r)$

Use the properties of addition to find the sum.

_____115_____ **19.** $4 + 105 + 6$

_____639_____ **20.** $575 + 25 + 39$

_____240_____ **21.** $78 + 140 + 22$

_____435_____ **22.** $41 + 235 + 159$

_____590_____ **23.** $318 + 90 + 182$

_____200_____ **24.** $33 + 124 + 17 + 26$

_____192_____ **25.** $92 + 85 + 15$

_____370_____ **26.** $92 + 47 + 23 + 208$

Properties of Multiplication

(Extra Practice; use after Section 3.3)

Name ___________________________________

Use the properties of multiplication to find each missing number. Identify the property illustrated.

Associative   **1.** $(18 \times 5) \times 13 = 18 \times (5 \times \underline{\,13\,})$

Identity   **2.** $987 \times \underline{\,1\,} = 987$

Commutative   **3.** $\underline{\,84\,} \times 48 = 48 \times 84$

Distributive   **4.** $5 \times (9 + 4) = (\underline{\,5\,} \times 9) + (\underline{\,5\,} \times 4)$

Associative   **5.** $375 \times (9 \times 12) = (375 \times \underline{\,9\,}) \times 12$

Zero Property   **6.** $\underline{\,0\,} \times 76 = 0$

Associative   **7.** $87 \times (\underline{\,248\,} \times 9) = (87 \times 248) \times 9$

Distributive   **8.** $5 \times (\underline{\,13\,} + 2) = (5 \times 13) + (5 \times 2)$

Commutative   **9.** $(123 \times 5) \times 47 = (5 \times \underline{\,123\,}) \times 47$

Zero Property   **10.** $(697 \times 0) \times (754 \times 27) = \underline{\,0\,}$

Match each property of multiplication with its definition.

A. Changing the order of the factors does not change the product.

B. The product of one and any number is that number.

C. The product of a factor and a sum is equal to the sum of the products.

D. Changing the grouping of the factors does not change the product.

E. The product of zero and any number is zero.

_D_ **11.** Associative

_A_ **12.** Commutative

_C_ **13.** Distributive

_B_ **14.** Identity

_E_ **15.** Zero

Applying the Distributive Property

You can use the Distributive Property to help you do math mentally. ①

Example 1: Multiply $2.07 × 7.
Answer:

$$7 × \$2.07 = 7 × (2.00 + 0.07)$$
$$= (7 × 2.00) + (7 × 0.07)$$
$$= 14.00 + 0.49$$
$$= \$14.49$$

Example 2: Multiply $1.97 × 7.
Answer:

$$7 × \$1.97 = 7 × (2.00 - 0.03)$$
$$= (7 × 2.00) - (7 × 0.03)$$
$$= 14.00 - 0.21$$
$$= \$13.79$$

Solve these problems using the Distributive Property.

_____$5.25_____ **1.** 5 × $1.05

_____$70.84_____ **2.** 7 × $10.12

_____$19.80_____ **3.** 6 × $3.30

_____$8.40_____ **4.** 3 × $2.80

_____$29.60_____ **5.** 8 × $3.70

_____$3.96_____ **6.** 4 × $0.99

_____$4.60_____ **7.** 5 × $0.92

_____$23.60_____ **8.** 8 × $2.95

_____$27.30_____ **9.** 7 × $3.90

_____$19.96_____ **10.** 2 × $9.98

Find the answer two different ways.

11. Suzanne is helping her mother put up a wallpaper border in her bedroom. The ceiling is high enough for the border to go above all the doors and windows. The room is 13 ft long and 10 ft wide. How many feet of border will be needed? (Draw a picture.)

$$2(l + w) = 2l + 2w$$
$$2(13 + 10) = 2 × 13 + 2 × 10$$
$$2 × 23 = 26 + 20$$
$$46 \text{ ft} = 46 \text{ ft}$$

Ask the students "which property of mathematics was used?" $2(l + w) = 2l + 2w$ is an application of the Distributive Property.

① Emphasize that the right side of the equation shows the thought process. Help the students understand when it is easier to use subtraction instead of addition. Since the Distributive Property was not defined for subtraction, the second example may need some explanation. Direct the students to compute 7 × $1.97 and then compare their work to the right side of the equation. Then, discuss why this mental math skill is helpful in everyday life.

Applying Linear Expressions

Name _______________________________________

Frieda runs a dog grooming service and has three people that work for her: Amy, Beth, and Cindy. When a dog comes in, Amy washes it, Beth trims the nails and coat, and then Cindy puts on the finishing touches (brushing, bows, nail polish, etc.). The girls get paid an hourly rate based on how long they have worked for Frieda, plus a set amount per dog. Amy gets $8 per hour, plus $5 per dog. Beth gets $7 per hour, plus $6 per dog. Cindy gets $9.50 per hour, plus $4.50 per dog.

Write an expression for each groomer's pay for a given day.
Use the variables h for hours worked and d for dogs groomed.

1. Amy

$8h + 5d$

2. Beth

$7h + 6d$

3. Cindy

$9.5h + 4.5d$

4. Assuming everyone works the same number of hours every day, combine the answers from 1–3 to get an expression for the amount Frieda will have to pay her groomers in a day. $24.5h + 15.5d$

5. If Frieda's shop is open Tuesday–Saturday every week, multiply to get an expression for the amount Frieda will have to pay her groomers in a week. Note that h and d still stand for the number of hours and dogs per day. $122.5h + 77.5d$

If a typical day sees everyone working for 8 hours grooming about 12 dogs, calculate the amount of daily pay for each groomer. Who gets paid the most?

6. Amy

$124

7. Beth

$128

8. Cindy

$130 (most)

9. If Frieda charges $50 per dog for grooming, how much will she have left for other expenses and profit after she pays her groomers for a week's work? (Hint: You need to figure out how many dogs they would groom in a week, and then how much money that would bring in. Then use your expression in exercise 5 to calculate the weekly groomer expenses.) $1090

$50 \times 12 \times 5 = 3000 brought in per week
$122.5 (8) + 77.5 (12) = 1910 groomer expenses
$1090 left

Multiplying and Dividing by Powers of Ten

(Extra Practice; use after Section 3.8)

Multiply or divide.

_____37.5_____ **1.** 3.75×10

_____95_____ **2.** 0.95×100

_____4800_____ **3.** 4.8×1000

_____429_____ **4.** 42.9×10

_____0.082_____ **5.** $8.2 \div 100$

_____0.964_____ **6.** $9.64 \div 10$

_____0.0736_____ **7.** $73.6 \div 1000$

_____0.0032_____ **8.** $0.32 \div 100$

_____3.25_____ **9.** 0.00325×1000

_____1817.2_____ **10.** 18.172×100

Write each power of ten in exponential form.

_____10^2_____ **11.** 100

_____10^4_____ **12.** 10,000

_____10^1_____ **13.** 10

_____10^3_____ **14.** 1000

_____10^6_____ **15.** 1,000,000

Multiply or divide by the given power of ten.

_____41,371_____ **16.** 41.371×10^3

_____0.07612_____ **17.** $761.2 \div 10^4$

_____0.00012_____ **18.** $0.012 \div 10^2$

_____400,160_____ **19.** 4.0016×10^5

_____0.000782_____ **20.** $782 \div 10^6$

Scientific Notation

Name ________________________________

Write in scientific notation.

_______3.6×10^3_______ **1.** 3600

_______4.28×10^4_______ **2.** 42,800

_______5.02×10^4_______ **3.** 50,200

_______5.6×10^6_______ **4.** 5,600,000

_______4.3×10^5_______ **5.** 430,000

_______6.2×10^8_______ **6.** 620,000,000

_______9×10^9_______ **7.** 9,000,000,000

_______8.92×10^{11}_______ **8.** 892,000,000,000

_______5.28×10^{13}_______ **9.** The star Sirius is 52,800,000,000,000 mi away. Write this distance in scientific notation.

_______6×10^{17}_______ **10.** The Milky Way Galaxy is 600,000,000,000,000,000 mi across. Write this distance in scientific notation.

Write in standard form.

_______460,000_______ **11.** 4.6×10^5

_______3,140,000_______ **12.** 3.14×10^6

_______12,040,000_______ **13.** 1.204×10^7

_______407,000,000_______ **14.** 4.07×10^8

_______5,062,000,000_______ **15.** 5.062×10^9

_______98,000,000,000_______ **16.** 9.8×10^{10}

_______726,800,000,000_______ **17.** 7.268×10^{11}

_______200,300,000,000_______ **18.** 2.003×10^{11}

19. The nearest star is 2.52×10^{13} mi away. Write this as a standard numeral.
25,200,000,000,000

20. The distance from the Earth to Pluto is approximately 3.6×10^9 mi. Write this distance as a standard numeral. 3,600,000,000

Applying Scientific Notation
(Mixed Practice; use after Section 3.8)

Calculate the answers to the following questions. Be sure your answers are all in proper scientific notation.

1. At a storage warehouse, each crate weighs 1.5×10^2 lb. If there are 5.2×10^3 crates, what is their combined weight? 7.8×10^5 lb

2. There are 5.6×10^2 workers in a processing plant, and each worker can fill 8.2 crates with fruit every day. What is the total number of crates filled by the workers each day? 4.592×10^3 crates

3. A factory can make 4.7×10^4 T-shirts per day, how many T-shirts will it make in 3.1×10^2 days? 1.46×10^7 T-shirts

4. If there are 40 people per square kilometer in a city, and the area of the city is 2.1×10^4 km^2, what is the total number of people in the city? 8.4×10^5 people

5. A spacecraft travels at a speed of 7.8×10^4 miles per day. How far will it travel in 1.5×10^2 days? 1.17×10^7 mi

6. All of the schools in a county banded together for a community service project. Each school planted 8.1×10^3 trees that year. If there are 1.01×10^2 schools in the county, how many trees did they plant altogether? 8.181×10^5 trees

7. Challenge question! If there are 6.2×10^4 students in the county, about how many trees did each student plant? (Hint: To find the answer, you need to *divide*.) 1.3×10^1 or 13 trees each

Scientific Notation with Negative Exponents

(Enrichment; use after Section 3.8)

Name _______________________

Scientific notation is a useful tool for calculations that involve very large numbers, such as distances in space. But it is also useful for very tiny numbers, as the size of a germ. To work with these numbers, you need to use negative exponents.

You should be familiar with the following pattern:

$$10^0 = 1; \; 10^1 = 10; \; 10^2 = 100; \; 10^3 = 1000; \; 10^4 = 10{,}000; \; 10^5 = 100{,}000; \text{ and so on.}$$

Negative exponents follow a similar pattern:

$$10^{-1} = \frac{1}{10} = 0.1; \; 10^{-2} = \frac{1}{100} = 0.01; \; 10^{-3} = \frac{1}{1000} = 0.001; \; 10^{-4} = \frac{1}{10{,}000} = 0.0001; \text{ and so on.}$$

So for instance, to change 0.000125 to scientific notation, you move the decimal point to the *right* to form a number between 1 and 10. Count the number of digits the decimal place was moved and write the number of places the decimal point moved to the right as the *negative* exponent of 10. $0.000125 = 1.25 \times 10^{-4}$.

Write in scientific notation.

3.92×10^{-5} **1.** 0.0000392

4.3×10^{-2} **2.** 0.043

8.45×10^{-8} **3.** 0.0000000845

7.837×10^{-1} **4.** 0.7837

Write in standard form.

0.006214   **5.** 6.214×10^{-3}

0.0000091   **6.** 9.1×10^{-6}

0.000893   **7.** 8.93×10^{-4}

0.03   **8.** 3×10^{-2}

Multiplying using scientific notation with negative exponents works the same as with positive exponents. You just need to keep in mind what you know about adding and subtracting integers.

9. A coronavirus particle measures around 120 nanometers. A nanometer is a *billionth* of a meter. So that is $120 \times \frac{1}{1{,}000{,}000{,}000}$ meters. Put that number in proper scientific notation. 1.2×10^{-7}

10. One cough from a sick person can contain as many as two hundred million coronavirus particles. (Yikes! Cover your mouth!) Put that number in proper scientific notation. 2×10^8

11. Imagine that for some odd reason we could lay all of those particles end to end. How many meters long would the line be? (Hint: multiply the number of particles by the size of each particle.) 2.4×10^1 or 24 m

Modern weather observation instruments use advanced technology. But with a little research, ingenuity, and some common materials, you can build your own devices to measure and even predict the weather.

PROCEDURES

Planning the Design

1. Conduct research on common weather observation instruments.

2. List five common weather parameters that can be measured with a homemade instrument. Include the name of the instrument.

3. With your group (or individually) draw and label a proposed design for 1 of the 5 instruments. Include a list of required materials.

Testing the Design

1. When approved by your teacher, construct the instrument according to your plan.

2. Use your completed instrument to record a weather observation.

Refining the Design

1. Use the results from your weather observation to determine what design changes you can make to improve the accuracy, ease of use, or utility of your instrument.

2. Draw and label a new design diagram.

3. Build a new instrument or modify the existing one to incorporate your revised design.

Retesting the Design

1. Use the "Observations" form to record readings from your instrument over the course of several days. Modify the days and times to fit your schedule. For comparison, record data from a local weather station or weather app in the appropriate place.

2. After several days, compare your data with local official observations and with weather patterns for your area.

3. Submit your observations, the original and revised designs, and a report of how effective you think your instrument was in recording the weather.

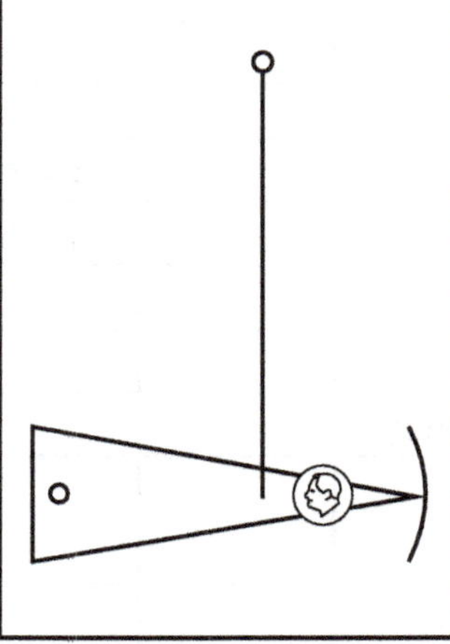

RECORD YOUR OBSERVATIONS

Weather Observations—Barometer						
Day	8:00	10:00	12:00	2:00	High/Low	Local Barometer
Monday						
Tuesday						
Wednesday						
Thursday						
Friday						
High/Low						

Weather Observations—Anemometer					
Day	8:00	10:00	12:00	2:00	High/Low
Monday					
Tuesday					
Wednesday					
Thursday					
Friday					
High/Low					

Weather Observations—Weathervane				
Day	8:00	10:00	12:00	2:00
Monday				
Tuesday				
Wednesday				
Thursday				
Friday				

Weather Observations—Hygrometer						
Day	8:00	10:00	12:00	2:00	High/Low	Relative Humidity
Monday						
Tuesday						
Wednesday						
Thursday						
Friday						
High/Low						

Weather Observations—Rain Gauge					
Day	8:00	10:00	12:00	2:00	High/Low
Monday					
Tuesday					
Wednesday					
Thursday					
Friday					
High/Low					

Name _______________________________

_____D_____ **1.** Round 4.675209 to the nearest thousandth. **[1.1]**

    A. 5   **B.** 4.68

    C. 4.7   **D.** not given

_____D_____ **2.** Subtract $27{,}602 - 8654$. **[1.2]**

    A. 18,858   **B.** 19,848

    C. 18,848   **D.** not given

_____C_____ **3.** Estimate $\$14.95 \times 6.38$ to the nearest dollar. **[1.3]**

    A. $8   **B.** $10

    C. $90   **D.** not given

_____B_____ **4.** Write 2^5 in standard form. **[1.5]**

    A. 64   **B.** 32

    C. 16   **D.** not given

_____B_____ **5.** Find $\sqrt{8 \times 50} \div 4$. **[1.7]**

    A. 40   **B.** 5

    C. 24   **D.** not given

_____A, C_____ **6.** Identify the set(s) to which the number −5 belongs. Choose all that apply. **[2.1]**

    A. $\mathbb{Z}$ (Integers)   **B.** $\mathbb{N}$ (Natural numbers)

    C. $\mathbb{Q}$ (Rational numbers)   **D.** $\mathbb{W}$ (Whole numbers)

_____B_____ **7.** Simplify $5 + |-3| \cdot 4$. **[2.2]**

    A. −7   **B.** 17

    C. 32   **D.** not given

_____B_____ **8.** Simplify $-21 + 15 + (-3)$. **[2.3]**

    A. 3   **B.** −9

    C. −39   **D.** not given

_____D_____ **9.** Simplify $(-8) \cdot (-9)$. **[2.5]**

    A. −17   **B.** 17

    C. −72   **D.** not given

_____C_____ **10.** Simplify $45 \div (-9)$. **[2.6]**

    A. 36   **B.** 5

    C. −5   **D.** not given

Divisibility

Name _______________________________

(Extra Practice; use after Section 4.1)

Is the first number divisible by the second? Write *yes* or *no*.

___yes___ **1.** 756; 3   ___no___ **2.** 829; 9

___yes___ **3.** 325; 5   ___no___ **4.** 654; 4

___no___ **5.** 412; 6   ___yes___ **6.** 5324; 4

___no___ **7.** 3728; 5   ___yes___ **8.** 6920; 2

___no___ **9.** 7133; 3   ___no___ **10.** 5890; 4

___yes___ **11.** 9204; 6   ___yes___ **12.** 86,418; 9

___yes___ **13.** 78,310; 10   ___no___ **14.** 309,214; 6

___yes___ **15.** 258,176; 4   ___no___ **16.** 514,298; 9

Classify each statement as true or false.

___True___ **17.** All even numbers are divisible by 2.

___False___ **18.** All odd numbers are divisible by 3.

___True___ **19.** If a number is divisible by 9, it is also divisible by 3.

___False___ **20.** If a number is divisible by 5, it is also divisible by 10.

___True___ **21.** If a number is divisible by 4, then it must be an even number.

___False___ **22.** If a number is divisible by 3, then it must be an odd number.

The Euclidean Algorithm
(Enrichment; use after Section 4.2)

Euclid was a Greek mathematician who lived about three hundred years before Christ. He discovered a method called the Euclidean Algorithm in which he used division to find the greatest common factor (GCF) of 2 numbers. To use the Euclidean Algorithm, follow these steps.

1. Divide the larger number by the smaller number.

2. Divide the divisor by the remainder.

3. Repeat the process until the remainder is 0.

4. The divisor of the last problem is the GCF.

Example: Find the GCF of 15 and 51.

Answer:

$$15 \overline{)51} \quad 3,\ 45,\ 6 \qquad 6\overline{)15} \quad 2,\ 12,\ 3 \qquad 3\overline{)6} \quad 2,\ 6,\ 0$$

GCF = 3

Find the greatest common factor using the Euclidean Algorithm. Show your work.

_____12_____ **1.** 12 and 24

_____12_____ **2.** 36 and 60

_____27_____ **3.** 54 and 81

_____7_____ **4.** 21 and 49

_____60_____ **5.** 600 and 180

_____6_____ **6.** 60 and 66

_____5_____ **7.** 35 and 25

_____3_____ **8.** 78 and 105

_____140_____ **9.** 420 and 700

_____21_____ **10.** 147 and 378

Practicing with the GCF and LCM

Name ____

Example: Find the GCF and LCM of 80 and 192.

Answer:

$80 = 2 \cdot 2 \cdot 2 \cdot 2 \cdot 5 = 2^4 \cdot 5$

$192 = 2 \cdot 2 \cdot 2 \cdot 2 \cdot 2 \cdot 2 \cdot 3 = 2^6 \cdot 3$

GCF (list the common prime factors) $= 2 \cdot 2 \cdot 2 \cdot 2 = 2^4 = 16$

LCM (list each prime factor with the highest exponent) $= 2^6 \cdot 3 \cdot 5 = 960$

To check your answer, multiply the given numbers, 80 and 192. Then, divide the product (15,360) by the GCF, 16. If you are correct, the quotient should be the LCM, 960.

Find the GCF and LCM of each pair of numbers given. Check your answers.

GCF = 16, LCM = 160   **1.** 32 and 80

GCF = 12, LCM = 168   **2.** 24 and 84

GCF = 24, LCM = 144   **3.** 48 and 72

GCF = 6, LCM = 432   **4.** 48 and 54

GCF = 4, LCM = 840   **5.** 56 and 60

GCF = 36, LCM = 216   **6.** 72 and 108

GCF = 4, LCM = 3192   **7.** 168 and 76

GCF = 1, LCM = 51,300   **8.** 76 and 675

Problem Solving Using the GCF and LCM

1. Maria is mounting insects on a display board for her science project. She needs to put the same number of insects in each row. What is the smallest number of insects Maria could mount if the following are all true?
   When she put them in 2 rows, she had 1 left over.
   When she put them in 3 rows, she had 1 left over.
   When she put them in 7 rows, she had none left over.   Maria had 49 insects. Our answer assumes that each row has more than 1 insect. Technically, student answers of "7" (1 insect per row) are correct.

2. Chris is buying a $60 gift for his mother. Chris has 4 brothers who may want to share the expenses equally. List all the possible contributions, depending on the number of brothers participating.   No brothers (Chris alone): $60; 1 brother and Chris: $30; 2 brothers and Chris: $20; 3 brothers and Chris: $15; 4 brothers and Chris: $12

3. At Sandyville Christian School, the third- and fourth-grade classes are going on a picnic. The teachers want to divide their classes into smaller teams of equal size. However, they do not want to mix third and fourth graders on the same team. If there are 24 third graders and 32 fourth graders, what is the largest possible number they could put on each team? How many teams would they have?   8 per team; 7 teams (3 for the 3rd grade and 4 for the 4th grade)

4. A school district is buying new furniture for some of the classrooms. Each classroom must have 28 desks. The furniture company gives a special discount for every 50 desks purchased. If the school district wants to furnish complete classrooms without having to store any desks and take full advantage of the special discount, what is the smallest number of desks that it should purchase? How many rooms can be furnished with new desks?   700 desks; 25 rooms

Terms Across and Down

Name ___________________________

For each description supply the appropriate term.

Across

   2. 2 numbers whose GCF is 1 (2 words)

   4. the relationship of 4, 6, 8, 10, . . . to 2

   8. the part of a fraction that tells the total number of parts or units

   9. the number that is a factor of every whole number

10. a number with more than 2 factors

11. a number greater than 1 that has exactly 2 factors

13. the GCF of 16 and 24

15. a fraction whose value is less than 1

16. the part of a fraction telling the amount being considered

19. the LCM of 4 and 6

Down

   1. a number that divides evenly into a given number

   3. 2 fractions with the same value

   5. the smallest prime number

   6. a fraction in which the numerator and denominator have a GCF of 1 (2 words)

   7. the smallest composite number

12. a fraction whose value is 1 or more

13. numbers that are divisible by 2

14. a numerical expression that represents part of a whole

17. division by 0

18. the ending digit of numbers divisible by both 10 and 5

Fill in the crossword puzzle based on the clues on the previous page.

Amazing Equivalent Fractions
(Mixed Practice; use after Section 4.4)

Name ______________________________

To help Oscar find his way through the maze, begin at START and compare the given fractions. Shade the octagons containing equivalent fractions. The shaded shapes will define a clear path to the END.

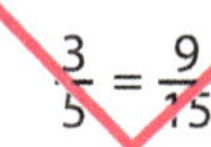

START

$\frac{2}{4} = \frac{1}{2}$ $\frac{3}{4} = \frac{1}{2}$ $1\frac{3}{4} = \frac{14}{8}$ $\frac{15}{12} = 1\frac{1}{3}$

$\frac{3}{4} = \frac{6}{8}$ $1\frac{3}{4} = \frac{7}{4}$ $\frac{1}{2} = \frac{2}{4}$ $\frac{81}{9} = 9\frac{1}{9}$

$\frac{3}{4} = \frac{9}{10}$ $\frac{10}{9} = 1\frac{1}{9}$ $\frac{1}{5} = \frac{3}{10}$ $\frac{2}{3} = \frac{8}{12}$

$\frac{9}{9} = \frac{1}{9}$ $1 = \frac{2}{3}$ $\frac{10}{40} = \frac{1}{4}$ $\frac{45}{9} = \frac{1}{5}$

$\frac{4}{6} = \frac{9}{12}$ $\frac{1}{5} = \frac{5}{25}$ $\frac{45}{25} = \frac{4}{2}$ $\frac{3}{2} = \frac{9}{6}$

$\frac{9}{15} = \frac{6}{10}$ $\frac{3}{7} = \frac{9}{21}$ $\frac{3}{2} = 1\frac{1}{2}$ $2\frac{3}{4} = \frac{11}{3}$

$\frac{2}{3} = \frac{12}{18}$ $\frac{3}{5} = \frac{4}{6}$ $\frac{45}{9} = 5$ $1\frac{2}{3} = \frac{19}{12}$

$\frac{6}{6} = 1$ $\frac{6}{6} = 6$ $\frac{3}{2} = 2\frac{1}{2}$ $\frac{25}{36} = \frac{5}{6}$

$\frac{3}{7} = \frac{2}{14}$ $2\frac{3}{4} = \frac{11}{4}$ $1\frac{1}{4} = \frac{10}{8}$ $1\frac{2}{3} = \frac{5}{3}$

$\frac{2}{3} = \frac{3}{2}$ $\frac{3}{5} = \frac{9}{15}$ $2 = \frac{2}{1}$ $\frac{3}{4} = \frac{9}{12}$

END

_____B_____ **1.** Choose the greatest decimal: 6.203, 6.21, 6.3. **[1.1]**

 A. 6.203
 B. 6.3
 C. 6.21
 D. none of these

_____B_____ **2.** Identify the most accurate estimate for 147,365 − 89,213. **[1.2]**

 A. 0
 B. 60,000
 C. 10,000
 D. 75,000

_____C_____ **3.** Between what two consecutive whole numbers does $\sqrt{456.998}$ lie? **[1.6]**

 A. 29 and 30
 B. 22 and 23
 C. 21 and 22
 D. none of these

_____C_____ **4.** List in increasing order: 0, 3, −3, 6, −6. **[2.2]**

 A. 0, −3, 3, −6, 6
 B. −3, −6, 0, 3, 6
 C. −6, −3, 0, 3, 6
 D. none of these

_____B_____ **5.** John Smith wrote checks totaling $310 and withdrew $105 in cash. His bank balance began at $402. What is the total in his account now? **[2.3]**

 A. $197
 B. −$13
 C. $607
 D. none of these

_____A_____ **6.** Simplify $(3 − 6) + 7(10 − 12)$. **[2.7]**

 A. −17
 B. −8
 C. 55
 D. none of these

_____D_____ **7.** Name the property illustrated by $3 \times 6 = 6 \times 3$. **[3.2]**

 A. Associative
 B. Identity
 C. Distributive
 D. none of these

_____ **8.** Simplify the following expression: $(18x + 14) + (2x - 6)$. **[3.5]**

     A. $20x + 8$
     B. $20x + 20$
     C. $16x + 8$
     D. none of these

_____ **9.** Simplify the following expression: $(m^3)^5$. **[3.7]**

     A. m^2
     B. m^8
     C. m^{15}
     D. none of these

_____ **10.** Change 234,000 and 32,000,000 each to scientific notation and multiply. **[3.8]**

     A. 7.488×10^{11}
     B. 74.8×10^{13}
     C. 7.488×10^{12}
     D. none of these

Adding Fractions
(Extra Practice; use after Section 5.1)

Name _______________________

Find each sum. Express all improper fractions as mixed numbers.

$3\frac{1}{2}$ **1.** $\frac{3}{2} + \frac{4}{2}$

$4\frac{1}{2}$ **2.** $\frac{11}{3} + \frac{5}{6}$

$-2\frac{11}{12}$ **3.** $-\frac{5}{3} + \left(-\frac{5}{4}\right)$

-1 **4.** $-\frac{3}{8} + \left(-\frac{5}{8}\right)$

5 **5.** $\frac{3}{2} + \frac{7}{2}$

$-\frac{1}{18}$ **6.** $-\frac{2}{9} + \frac{1}{6}$

$\frac{6}{35}$ **7.** $\frac{3}{5} + \left(-\frac{3}{7}\right)$

$\frac{7}{8}$ **8.** $\frac{1}{2} + \frac{1}{4} + \frac{1}{8}$

$-\frac{7}{24}$ **9.** $-\frac{2}{3} + \frac{3}{8}$

$\frac{2}{3}$ **10.** $\frac{13}{21} + \frac{5}{21} + \left(-\frac{4}{21}\right)$

$-\frac{11}{15}$ **11.** $-\frac{7}{12} + \left(-\frac{3}{20}\right)$

$\frac{1}{4}$ **12.** $-\frac{2}{3} + \frac{11}{12}$

$\frac{22}{35}$ **13.** $\frac{11}{30} + \frac{11}{42}$

$1\frac{7}{18}$ **14.** $\frac{8}{9} + \frac{5}{6} + \left(-\frac{1}{3}\right)$

Adding and Subtracting Fractions

Solve each problem. Then compare each answer with the one next to it by writing >, <, or = between the answers.

1. $\dfrac{3}{4}$ $+\dfrac{2}{8}$ = 1 $\quad<\quad$ $\dfrac{5}{6}$ $+\dfrac{1}{4}$ = $1\dfrac{1}{12}$

2. $\dfrac{2}{3}$ $+\dfrac{5}{6}$ = $1\dfrac{1}{2}$ $\quad>\quad$ $\dfrac{9}{11}$ $+\dfrac{1}{2}$ = $1\dfrac{7}{22}$

3. $\dfrac{8}{15}$ $-\dfrac{1}{2}$ = $\dfrac{1}{30}$ $\quad<\quad$ $\dfrac{9}{12}$ $-\dfrac{2}{3}$ = $\dfrac{1}{12}$

4. $\dfrac{2}{5}$ $+\dfrac{3}{4}$ = $1\dfrac{3}{20}$ $\quad<\quad$ $\dfrac{5}{7}$ $+\dfrac{2}{3}$ = $1\dfrac{8}{21}$

5. $\dfrac{7}{8}$ $-\dfrac{1}{2}$ = $\dfrac{3}{8}$ $\quad>\quad$ $\dfrac{8}{9}$ $-\dfrac{2}{3}$ = $\dfrac{2}{9}$

6. $\dfrac{7}{12}$ $+\dfrac{1}{2}$ = $1\dfrac{1}{12}$ $\quad<\quad$ $\dfrac{3}{5}$ $+\dfrac{2}{3}$ = $1\dfrac{4}{15}$

7. $\dfrac{7}{8}$ $-\dfrac{1}{4}$ = $\dfrac{5}{8}$ $\quad>\quad$ $\dfrac{5}{6}$ $-\dfrac{1}{3}$ = $\dfrac{1}{2}$

8. $\dfrac{6}{11}$ $-\dfrac{1}{3}$ = $\dfrac{7}{33}$ $\quad<\quad$ $\dfrac{7}{15}$ $-\dfrac{1}{4}$ = $\dfrac{13}{60}$

9. $\dfrac{8}{12}$ $+\dfrac{5}{6}$ = $1\dfrac{1}{2}$ $\quad=\quad$ $\dfrac{6}{9}$ $+\dfrac{10}{12}$ = $1\dfrac{1}{2}$

10. $\dfrac{16}{21}$ $-\dfrac{4}{7}$ = $\dfrac{4}{21}$ $\quad<\quad$ $\dfrac{19}{25}$ $-\dfrac{8}{15}$ = $\dfrac{17}{75}$

 Fundamentals of Math

Mixed Numbers Applications

Name _________________________________

Solve each problem.

1. Jasmine's mother is making matching dresses for Jasmine and her sister. If Jasmine's dress takes $3\frac{3}{8}$ yd of material and her sister's dress takes $2\frac{3}{4}$ yd, how much material should they purchase? $6\frac{1}{8}$ yd

2. The dresses call for $1\frac{2}{3}$ yd and $2\frac{1}{8}$ yd of trim, respectively. How much trim do they need? $3\frac{19}{24}$ yd

3. Jasmine finds a $4\frac{1}{4}$ yd long remnant that is identical to the trim needed for the dresses. After sewing the trim on the matching dresses, will Jasmine have enough trim left over to make a doll's dress that requires 1 ft ($\frac{1}{3}$ yd) of trim?

   Yes; $4\frac{1}{4} - 3\frac{19}{24} = \frac{17}{4} - \frac{91}{24} = \frac{102}{24} - \frac{91}{24} = \frac{11}{24}$ yd will be left, and $\frac{11}{24} - \frac{1}{3} = \frac{11}{24} - \frac{8}{24} = \frac{3}{24}$.

4. Hannah and her mother were baking four kinds of Christmas cookies. The recipes called for the following amounts of flour: $1\frac{1}{2}$ c, 2 c, $1\frac{1}{4}$ c, and $1\frac{2}{3}$ c. How much flour did they need? $6\frac{5}{12}$ c

5. The cookie recipes called for the following amounts of sugar: $\frac{3}{4}$ c, $1\frac{1}{3}$ c, $\frac{1}{2}$ c, and $\frac{2}{3}$ c. How much sugar did they need? $3\frac{1}{4}$ c

6. Ladies at Northside Christian School baked cookies for the annual Christmas program reception. Four ladies participated and baked $3\frac{1}{2}$ dozen, $5\frac{1}{4}$ dozen, $2\frac{3}{4}$ dozen, and 4 dozen cookies. How many cookies were baked for the reception? $15\frac{1}{2}$ dozen

7. Max worked $5\frac{1}{2}$ hr, $4\frac{3}{4}$ hr, $2\frac{1}{2}$ hr, $3\frac{1}{4}$ hr, and 3 hr this week. How many total hours did he work? 19 hr

8. Mr. DeYoung has 8 cows that he milks at his small farm. The gallons of milk they gave on July 1 were $3\frac{3}{8}$, $4\frac{3}{4}$, $3\frac{3}{4}$, $4\frac{1}{2}$, $4\frac{1}{4}$, $3\frac{3}{8}$, $4\frac{1}{8}$, and $4\frac{1}{8}$. How many gallons did Mr. DeYoung have to sell that day? If he had $35\frac{3}{8}$ gal to sell the day before, how much more or less did he have on July 1? $32\frac{1}{4}$ gal; had $3\frac{1}{8}$ gal less

9. A turtle race lasted for 10 min. At the end of the designated time, prizes were awarded to the owners of the 3 turtles that walked the farthest in the desired direction. The first-place turtle walked $2\frac{1}{4}$ ft, the second-place turtle walked $1\frac{3}{4}$ ft, and the third-place turtle walked $1\frac{1}{2}$ ft. How much farther did the first-place turtle walk than the second-place turtle? How much farther did the second-place turtle walk than the third-place turtle? $\frac{1}{2}$ ft; $\frac{1}{4}$ ft

10. Silas owns an orchard in Pennsylvania. Last year he sold $4\frac{1}{4}$ tn of grapes, $15\frac{3}{4}$ tn of apples, $5\frac{1}{2}$ tn of pears, and $12\frac{1}{3}$ tn of peaches. How many tons of fruit did he sell? $37\frac{5}{6}$ tn

Visualizing the Multiplication of Fractions

(Mixed Practice; use with Section 5.4)

Name _______________________

You can use shaded rectangles to depict multiplication of fractions. Think of the multiplication symbol as the word *of*. Study the following example.

Example: $\frac{1}{4} \times \frac{2}{3} \rightarrow \frac{1}{4}$ of $\frac{2}{3}$

Answer:

    1. The denominators are 4 and 3. Draw a 4×3 rectangle.

    2. Shade in $\frac{2}{3}$ of the rectangle (in this case, the bottom 2 rows).

    3. Darken $\frac{1}{4}$ of the shaded region.

    4. Determine what fractional part of the whole is represented by the darkened region. [2 parts of the 12 equal regions are darkened, or $\frac{2}{12}$ of the rectangle. Then simplify: $\frac{2}{12} = \frac{1}{6}$.]

Solve by drawing rectangles on the graph paper provided and using the procedure explained above.

1. $\frac{1}{2} \times \frac{7}{8}$ $\frac{7}{16}$

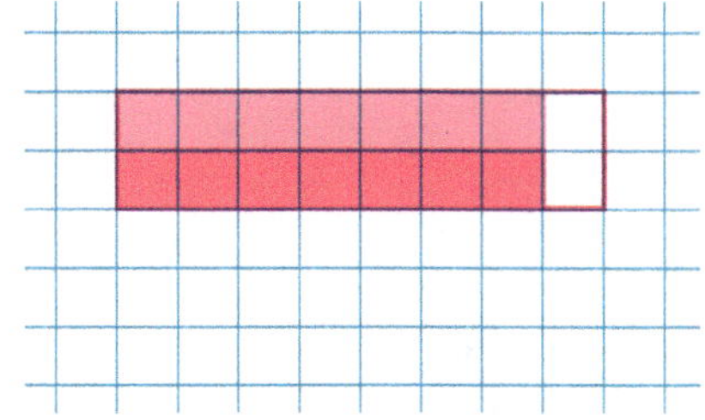

2. $\frac{1}{3} \times \frac{5}{6}$ $\frac{5}{18}$

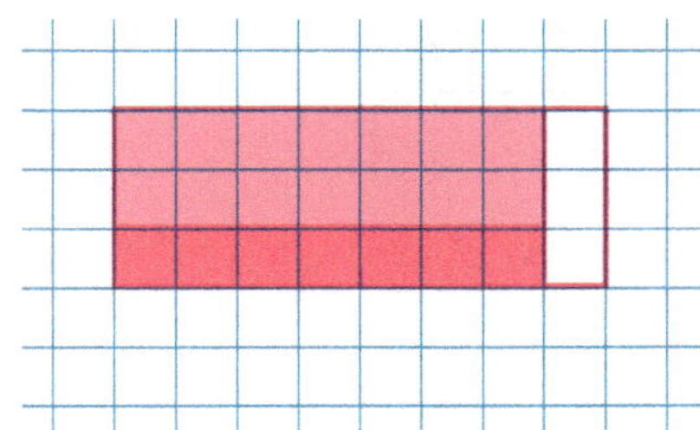

3. $\frac{1}{4} \times \frac{3}{5}$ $\frac{3}{20}$

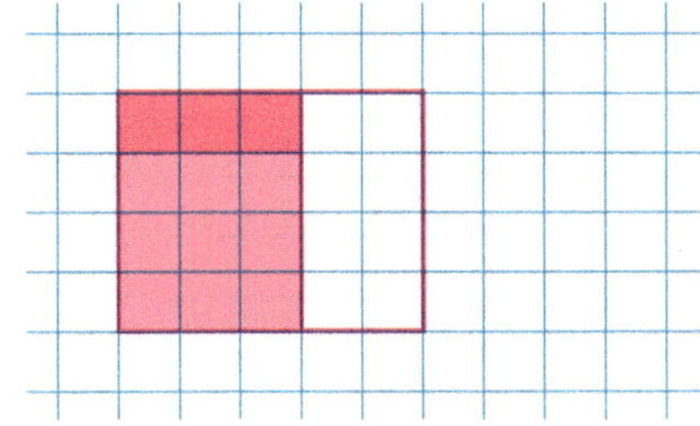

4. $\frac{3}{4} \times \frac{3}{8}$ $\frac{9}{32}$

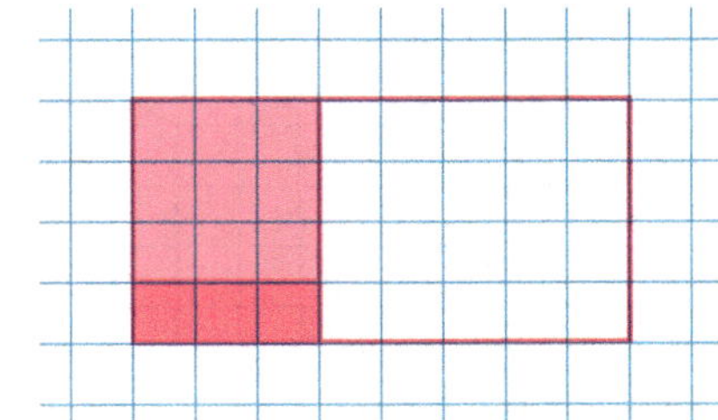

Visualizing the Division of Fractions

(Mixed Practice; use with Section 5.5)

You can use shaded rectangles to depict division of fractions. It is slightly more complicated, but it can still help you to visualize what is going on. Study the following example.

Example 1: $\frac{2}{3} \div \frac{1}{6}$

   Ask: [How many $\frac{1}{6}$s are there in $\frac{2}{3}$?]

Answer:

  1. The denominators are 3 and 6. Draw two 3 × 6 rectangles.

  2. Shade in $\frac{2}{3}$ of the first rectangle and $\frac{1}{6}$ of the other.

  3. Look at the second rectangle. How many squares are shaded? (3 squares are shaded.)

  4. Look back at the first rectangle. How many sets of 3 squares are shaded? (4 sets of 3 squares are shaded.) $\frac{2}{3} \div \frac{1}{6} = 4$

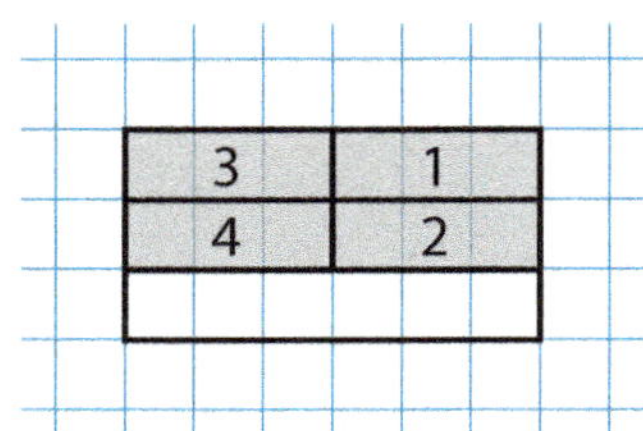

It works even when the answer is not a whole number!

Example 2: $\frac{3}{4} \div \frac{1}{2}$

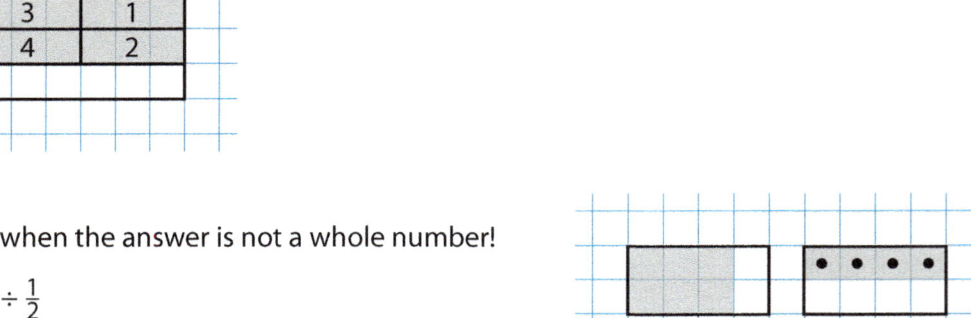

   Ask: How many $\frac{1}{2}$s are there in $\frac{3}{4}$?

Answer:

   $\frac{3}{4} \div \frac{1}{2} = 1\frac{1}{2}$ or $\frac{3}{2}$

Solve by drawing rectangles on the graph paper provided and using the procedure explained above.

  1. $\frac{3}{5} \div \frac{1}{10}$ 6

  2. $\frac{3}{5} \div \frac{2}{7}$ $\frac{21}{10}$ or $2\frac{1}{10}$

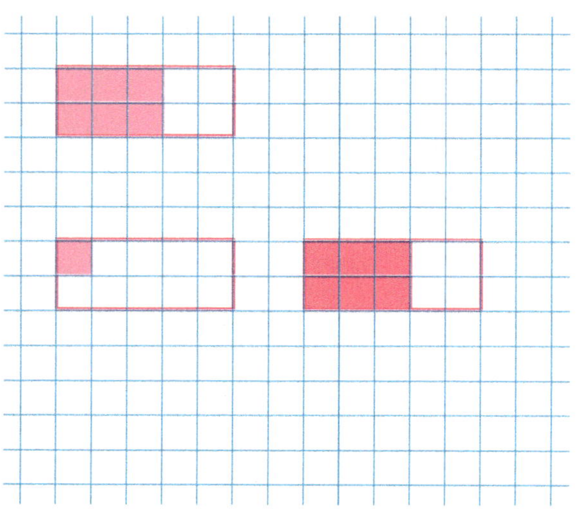

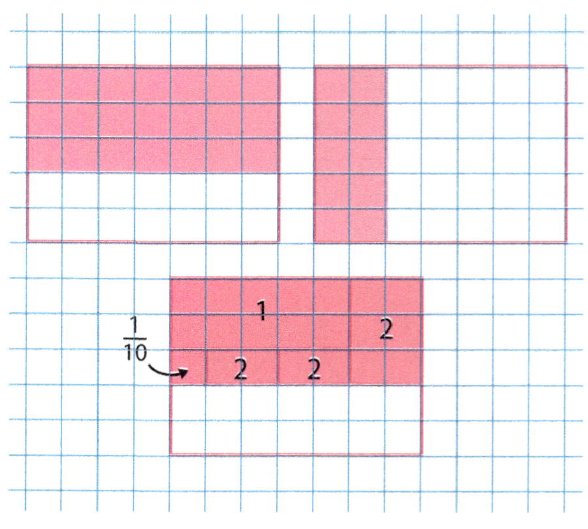

Fundamentals of Math

Mixed Numbers Applications

Name ________________________

1. Josie cut a $15\frac{3}{4}$ in. long ribbon into pieces that were $1\frac{1}{8}$ in. long. How many pieces of ribbon did she get?   14 pieces

2. Find the cost of $5\frac{1}{2}$ lb of hamburger if it sells for \$2 per pound.   \$11

3. If $\frac{2}{3}$ of a class of 24 students has brought in money for a field trip, how many students still need to bring in their money?   8 students

4. A gas station has 18 unleaded-gas pumps and 8 diesel pumps. If $10\frac{3}{4}$ gal of unleaded and $6\frac{2}{3}$ gal of diesel can be pumped per minute on each pump, how many gallons of gas can be pumped in 1 min? $246\frac{5}{6}$ gal

5. Mrs. Ward has $25\frac{3}{4}$ yd of material to make cheerleading skirts. If there are 6 girls on the squad, how much material is available for each skirt? $4\frac{7}{24}$ yd

Name _______________________________

_____ **1.** Round 0.2674 to the nearest thousandth. **[1.1]**

    A. 0.267
    B. 0.27
    C. 0.268
    D. none of these

_____ **2.** Subtract $1002.062 - 589.276$. **[1.2]**

    A. 412.786
    B. 413.786
    C. 412.886
    D. none of these

_____ **3.** Find $\sqrt{3600} + \sqrt{144} \div 2$. **[1.7]**

    A. 306
    B. 36
    C. 606
    D. none of these

_____ **4.** Consider the following sets. What set represents the elements that A and C have in common? $A = \{2, 4, 6, 8, \ldots\}$, $B = \{1, 3, 5, 7, 9, \ldots\}$, $C = \{-2, -1, 0, 1, 2\}$ **[2.1]**

    A. $\{1\}$
    B. $\{2\}$
    C. $\varnothing$
    D. not enough information

_____ **5.** A game show contestant had a score of −75. He missed another question, which means that he lost 200 more points. What is his current score? **[2.4]**

    A. −125
    B. −175
    C. −275
    D. none of these

_____ **6.** The price of a certain stock went down $2.50 on Monday, up $5 on Tuesday, down $3.50 on Wednesday, down $1.50 on Thursday, and up $2 on Friday. What was the average daily change in price? **[2.6]**

    A. −$0.50
    B. −$0.10
    C. $0.10
    D. none of these

_____ **7.** Simplify $2 \cdot 5 - (-3)$. **[2.7]**

    A. 4
    B. 16
    C. 13
    D. none of these

_____ **8.** Name the property illustrated: $(32 + 64) + 19 = 32 + (64 + 19)$. **[3.2]**

    A. Commutative
    B. Distributive
    C. Associative
    D. none of these

9. Which of the following is not equivalent to the expression $48n + 24$? **[3.4]**

A. $4(12n + 6)$
B. $3(16n + 8)$
C. $16(3n + 2)$
D. all are equivalent

10. Simplify the following expression: $x^5 \cdot x^4$. **[3.7]**

A. x^{20}
B. x^9
C. x
D. none of these

11. Write 7,540,000 in scientific notation. **[3.8**]

A. 7.54×10^6
B. 7.54×10^4
C. 75.4×10^5
D. none of these

12. To find all the primes that are factors of 4159, it is necessary to check which of the following? **[4.1]**

A. primes less than 4159
B. primes up to $\sqrt{4159}$
C. primes less than or equal to $\frac{4159}{2}$
D. none of these

13. Find the least common multiple of 12, 20, and 36. **[4.3]**

A. 180
B. 720
C. 360
D. none of these

14. Which fraction is not equivalent to the others? $\frac{9}{21}, \frac{15}{35}, \frac{21}{49}, \frac{24}{56}$ **[4.4]**

A. $\frac{9}{21}$
B. $\frac{24}{56}$
C. $\frac{15}{35}$
D. All are equivalent.

15. Express 0.44 as a lowest-term ratio of integers. **[4.5]**

A. $\frac{44}{100}$
B. $\frac{22}{50}$
C. $\frac{4.4}{10}$
D. none of these

Solving Addition and Subtraction Equations

(Extra Practice; use after Section 6.2)

Name _______________________________

To solve an equation, use the inverse operation. Make sure that the same number is added to or subtracted from both sides of the equation.

Solve each equation. Show your work and check your answers.

_____ $x = 15$ _____ **1.** $x + 4 = 19$

_____ $c = 9$ _____ **2.** $c + 5 = 3 \cdot 4 + 2$

_____ $c = 13$ _____ **3.** $c - 4 = 9$

_____ $y = 12$ _____ **4.** $y - 4 = 2 + 4 \cdot 2 - 6 \div 3$

_____ $y = 18$ _____ **5.** $y + 5 = 23$

_____ $a = 2$ _____ **6.** $a + 2 = (12 + 16) \div 4 - 3$

_____ $z = 26$ _____ **7.** $z + 2 = 28$

_____ $z = 42$ _____ **8.** $z - 9 = 3(4 + 7)$

_____ $a = 21$ _____ **9.** $a - 4 = 17$

_____ $x = 7$ _____ **10.** $x + 12 = 3 \cdot 8 - 25 \div 5$

_____ $b = 28$ _____ **11.** $b + 9 = 37$

_____ $e = 29$ _____ **12.** $e - (4 + 3 \cdot 2) = \frac{8}{2} + 3 \cdot 5$

_____ $e = 66$ _____ **13.** $e - 19 = 47$

_____ $h = 16$ _____ **14.** $h - 2(3 + 4) = 14 - 4 \cdot 3$

_____ $f = 98$ _____ **15.** $f - 21 = 77$

Solving Multiplication and Division Equations

(Extra Practice; use after Section 6.3)

To solve an equation use the inverse operation. Make sure that the same operation is applied to both sides of the equation.

Solve each problem. Show your work and check your answers.

_____$n = 13$_____ **1.** $7n = 91$ _____$z = -13$_____ **2.** $4z = -52$

_____$a = 15$_____ **3.** $9a = 135$ _____$b = 16$_____ **4.** $3b = 48$

_____$c = 112$_____ **5.** $\dfrac{c}{-4} = -28$ _____$c = 84$_____ **6.** $\dfrac{c}{6} = 14$

_____$h = 225$_____ **7.** $\dfrac{h}{5} = 45$ _____$d = -600$_____ **8.** $-\dfrac{d}{5} = 120$

_____$m = 14$_____ **9.** $8m = 112$ _____$x = -4$_____ **10.** $2x = -5 + (-3)$

_____$r = 189$_____ **11.** $\dfrac{r}{9} = 21$ _____$x = 9$_____ **12.** $3x = 33 - 6$

_____$s = 119$_____ **13.** $\dfrac{s}{7} = 17$ _____$x = 64$_____ **14.** $\dfrac{x}{4} = 21 - 5$

_____$m = 25$_____ **15.** $-18m = -450$

Solving Two-Step Equations

(Extra Practice; use after Section 6.4)

Name _______________________________________

Solve the following equations.

___ $n = -11$ ___ **1.** $6 + 2n = -16$

___ $x = 24$ ___ **2.** $2x - 10 = 38$

___ $x = 3$ ___ **3.** $3x - 12 = -3$

___ $x = 13$ ___ **4.** $-4x + 2 = -50$

___ $b = -32$ ___ **5.** $\dfrac{b}{-4} - 3 = 5$

___ $m = -30$ ___ **6.** $\dfrac{m}{3} - 8 = -18$

___ $y = -9$ ___ **7.** $-7y + 7 = 70$

___ $j = 8$ ___ **8.** $-j + 14 = 6$

___ $v = -6$ ___ **9.** $8v - 16 = -64$

___ $z = 3$ ___ **10.** $2z - 9 = -3$

___ $g = 15$ ___ **11.** $5 + \dfrac{g}{-5} = 2$

___ $w = 1$ ___ **12.** $w - (-4 - 7) = 12$

___ $m = 2$ ___ **13.** $\dfrac{m}{2} - 15 = -14$

___ $h = -63$ ___ **14.** $\dfrac{h}{-3} - 6 = 15$

___ $t = -7$ ___ **15.** $5t - 4 = -39$

Writing and Solving Equations

(Mixed Practice 6.1, 6.4; use after Section 6.4)

Write an equation for each problem and solve it.

1. The sum of a number and 17 is 52. What is the number? $n + 17 = 52, n = 35$

2. When a number is multiplied by 15, the product is 105. What is the number?
   $15n = 105, n = 7$

3. When a number is divided by 8, the quotient is 12. What is the number? $\frac{n}{8} = 12, n = 96$

4. If 25 is subtracted from a number, the difference is 9. What is the number?
   $n - 25 = 9, n = 34$

5. When a number is multiplied by 8 and 6 is added to the product, the total is 30.
   What is the number? $8n + 6 = 30, n = 3$

6. When 5 is subtracted from the quotient of a number and 2, the difference is 4.
   What is the number? $\frac{n}{2} - 5 = 4, n = 18$

7. David earns \$9 an hour. If David earns \$405, how many hours did he work?
   $9h = 405, h = 45$ hours

8. Sue buys a used car that costs \$8750. She pays \$2450 down on the car. How much
   more does she owe? $n + 2450 = 8750; n = \$6300$

9. Andrew put aside \$45 for his vacation. How much does he need to put aside each
   month in order to have \$317 at the end of 8 months? $8m + 45 = 317, m = \$34$

10. Ashley bought 5 cans of green beans. She used a coupon for \$0.45 off 1 can of
    green beans. The green beans cost \$3.80 altogether. What was the cost of each
    can of green beans without the coupon? $5c - 0.45 = 3.80, c = \$0.85$

11. Read Deuteronomy 2:14 and Joshua 14:6–12. At this point, "the land [of Canaan] had
    rest from war" (Joshua 14:15). How long does it appear that Joshua had waged war
    from Jericho to this point? 40 (Caleb's age at Kadesh-barnea) + 38 (time from Kadesh-
    barnea until they entered Canaan) + x (length of war) = 85 (Caleb's current age);
    $x = 7$ years

Fundamentals of Math

Name _______________________________

Solve and check.

1. $2(x + 3) = 4$ $x = -1$

2. $5(y - 4) = 20$ $y = 8$

3. $4(a + 7) = 16$ $a = -3$

4. $8(m - 1) = 24$ $m = 4$

5. $7(3d - 1) = 77$ $d = 4$

6. $3(2m + 3) = 21$ $m = 2$

7. $\frac{1}{2}(x + 7) = 2$ $m = -3$

8. $\frac{1}{3}(b - 9) = -1$ $b = 6$

9. $\frac{3}{2}(x + 4) = 9$ $x = 2$

10. $\frac{5}{4}(z + 3) = -10$ $z = -11$

11. $0.6(n - 5) = 3$ $n = 10$

12. $0.5(y + 10) = 8$ $y = 6$

13. $0.8(g + 3) = 4$ $g = 2$

14. $1.5(x - 7) = 9$ $x = 13$

15. $0.75(3a - 5) = 12$ $a = 7$

16. $\frac{1}{2}(3c + 1) = -4$ $c = -3$

Name _______________________________

V is voltage in volts; *I* is current in amps; *R* is resistance in ohms (Ω)

Can you connect a 9-volt battery directly to a 20-milliamp LED? Trial and error is not a good way to answer this question! Ohm's Law states a fundamental relationship between voltage, current, and resistance in electrical circuits. In this project you will use Ohm's Law to determine the appropriate resistor to use so that a 9-volt battery can effectively power an LED rated for a maximum current of 20 milliamps. A resistor that is too weak will shorten the life of the LED, while a resistor that is too strong will result in no light or a dim light. Connecting the 9-volt battery to the LED without a resistor will destroy the LED.

PROCEDURES

Planning the Design

1. Conduct research on Ohm's Law and simple electric circuits.

2. Use the properties of equality to solve Ohm's Law for current and for resistance.

3. With your group (or individually), draw and label a proposed design for an electric circuit that includes a 9-volt battery, an LED, and a resistor.

Testing the Design

1. When approved by your teacher, construct the circuit according to your plan.

2. Use a meter to check the voltage, current, and resistance in your circuit.

Refining the Design

1. Use the results from your test to determine what design changes you can make to improve your circuit.

2. Draw and label a new design diagram.

3. Build a new circuit that incorporates your revised design.

Retesting the Design

1. Use a meter to check the voltage, current, and resistance in your redesigned circuit.

2. Compare the actual readings from the meters to what you would expect based on Ohm's Law.

3 Submit your original and revised designs along with the expected and actual readings for voltage, current, and resistance for the revised circuit.

Solving Two-Step Inequalities
(Extra Practice; use after Section 6.8)

Name ______________________________

Solve each inequality.

_____ $n \neq -8$ _____ **1.** $7 + 3n \neq -17$

_____ $a \geq 72$ _____ **2.** $\frac{a}{12} - 9 \geq -3$

_____ $b < 81$ _____ **3.** $11 - \frac{b}{9} > 2$

_____ $z \geq 7$ _____ **4.** $-5(2z - 9) \leq -25$

_____ $x > \frac{1}{3}$ _____ **5.** $11 < 15x + 6$

_____ $y \leq 4$ _____ **6.** $\frac{2}{7}(14y - 35) \leq 6$

_____ $w \leq -12$ _____ **7.** $-\frac{w}{2.4} - 33 \geq -28$

_____ $d < 3$ _____ **8.** $1\frac{9}{20} > \frac{d}{4} - \left(-\frac{7}{10}\right)$

_____ $p > -1$ _____ **9.** $-24 > -8(2p + 5)$

_____ $v \neq -\frac{5}{8}$ _____ **10.** $17 + 16v \neq 7$

Determine whether the given value is a solution for each inequality.

_____ no _____ **11.** $12 - 5s < -3; s = 3$

_____ yes _____ **12.** $\frac{f}{-8} - 9 \leq -12; f = 32$

_____ yes _____ **13.** $\frac{3}{5}r + 4 \geq 10; r = 15$

_____ no _____ **14.** $215j - 743 \neq 117; j = 4$

Writing and Solving Inequalities

Write an inequality for each problem and solve it.

Four friends are doing after-school jobs to earn money for their summer mission trip.

1. Olivia sells at most 8 dozen chocolate chip cookies for $0.50 each during Late Stay each week. She pays her mom back for the ingredients out of the money she earns. The ingredients cost $1.25 per dozen. If she sells everything she bakes (without eating any!), what is the most that she can make in a week? $m \le 96(0.5) - 1.25(8); m \le \$38$

2. Charlotte is tutoring her sixth-grade friend Raelynn. On the first 2 tests of the semester, Raelynn got a 75 and a 68, but since Charlotte started working with Raelynn, she has gotten an 82 and a 90 on the last 2 tests. What does she need on the fifth test to average at least an 80 for all 5 tests in the semester?
$(75 + 68 + 82 + 90 + t) \div 5 \ge 80; t \ge 85$

3. Clay is walking dogs. He charges $6 for a half-hour walk. Currently, he has a dog that he walks every day at 4:00 and another at 5:00. But he realized he could walk more than one dog at a time—and he has an open slot at 4:30. How many more dogs would he need to add to his routine to be earning at least $54 per day?
$12 + 6d \ge 54; d \ge 7$ dogs

4. Ian is making a deal with his neighbor to blow the snow off his driveway every time it snows. The neighbor has offered Ian a choice of 2 options. Option 1: $120 per month, to cover unlimited snow events. Option 2: $15 for each snow event. How many snow days would there need to be in a month for the $120 to be a better deal for Ian?
$15s < 120; s < 8$ days

                    Name _______________________________

<u>D</u> **1.** Place the following decimal numbers in decreasing order:
23.6089, 23.$\overline{6}$, 23.712, 23.$\overline{65}$, 23.499, 23.481. **[1.1]**

    A. 23.6089, 23.712, 23.$\overline{6}$, 23.$\overline{65}$, 23.499, 23.481
    B. 23.712, 23.6089, 23.$\overline{65}$, 23.$\overline{6}$, 23.499, 23.481
    C. 23.712, 23.$\overline{65}$, 23.$\overline{6}$, 23.6089, 23.499, 23.481
    D. none of these

<u>B</u> **2.** Estimate $2.89\overline{)164.82}$. **[1.4]**

    A. 5000
    B. 50
    C. 500
    D. none of these

<u>A</u> **3.** Tell whether the following set is finite, infinite, or empty:
{even whole numbers less than 1000}. **[2.1]**

    A. finite
    B. infinite
    C. empty
    D. not enough information

<u>A</u> **4.** Simplify $|-36 \div 9| + 4$. **[2.2]**

    A. 8
    B. 0
    C. −8
    D. none of these

<u>B</u> **5.** Simplify $(-2)^3 + (-3)^2 - 4^2$. **[2.7]**

    A. 1
    B. −15
    C. 17
    D. none of these

<u>B</u> **6.** Translate the following phrase into an algebraic expression:
"6 less than the product of 4 and a number." **[3.1]**

    A. $6 - 4n$
    B. $4n - 6$
    C. $(6 - 4)n$
    D. none of these

<u>C</u> **7.** Simplify the following expression: $(8x + 7) - (2x - 3)$. **[3.6]**

    A. $6x + 4$
    B. $6x - 4$
    C. $6x + 10$
    D. none of these

<u>C</u> **8.** Multiply $(51.7 \times 10^6) \times (9.8 \times 10^4)$. Give your answer in scientific notation. **[3.8]**

    A. 506.66×10^{10}
    B. 5.0666×10^8
    C. 5.0666×10^{12}
    D. none of these

_____B_____ **9.** Which list contains only prime numbers? **[4.1]**

    A. 17, 19, 27
    B. 11, 41, 47
    C. 13, 23, 91
    D. none of these

_____C_____ **10.** Find the greatest common factor of 124 and 98. **[4.2]**

    A. 1
    B. 6
    C. 2
    D. none of these

_____D_____ **11.** List the following fractions in decreasing order: $\frac{5}{12}, \frac{2}{5}, \frac{7}{15}, \frac{8}{17}$. **[4.6]**

    A. $\frac{7}{15}, \frac{5}{12}, \frac{2}{5}, \frac{8}{17}$
    B. $\frac{8}{17}, \frac{5}{12}, \frac{7}{15}, \frac{2}{5}$
    C. $\frac{5}{12}, \frac{7}{15}, \frac{8}{17}, \frac{2}{5}$
    D. none of these

_____B_____ **12.** Add $\frac{3}{16} + \frac{7}{8} + \frac{1}{4}$. **[5.1]**

    A. $\frac{11}{28}$
    B. $1\frac{5}{16}$
    C. $2\frac{1}{16}$
    D. none of these

_____A_____ **13.** Subtract $6\frac{1}{3} - 4\frac{4}{5}$. **[5.3]**

    A. $1\frac{8}{15}$
    B. $1\frac{1}{5}$
    C. $2\frac{8}{15}$
    D. none of these

_____C_____ **14.** Multiply $\frac{3}{8} \times \frac{5}{9} \times \frac{4}{25}$. **[5.4]**

    A. $\frac{1}{10}$
    B. $\frac{3}{20}$
    C. $\frac{1}{30}$
    D. none of these

_____B_____ **15.** Find the quotient of $\frac{14}{15} \div \frac{3}{5}$. Express the answer in lowest terms. **[5.5]**

    A. $1\frac{25}{45}$
    B. $1\frac{5}{9}$
    C. $\frac{14}{25}$
    D. none of these

Writing and Simplifying Ratios
(Extra Practice; use after Section 7.1)

Name _______________________________

Use the favorite sport survey given below. Write the ratios as lowest-term fractions.

Favorite Sport Survey Chart	
Sport	**Number of Responses**
softball	40
basketball	75
soccer	150
football	35

$\dfrac{40}{300} = \dfrac{2}{15}$ **1.** softball to total

$\dfrac{40}{150} = \dfrac{4}{15}$ **2.** softball to soccer

$\dfrac{75}{150} = \dfrac{1}{2}$ **3.** basketball to soccer

$\dfrac{35}{150} = \dfrac{7}{30}$ **4.** football to soccer

$\dfrac{35}{40} = \dfrac{7}{8}$ **5.** football to softball

$\dfrac{150}{300} = \dfrac{1}{2}$ **6.** soccer to total

Refer to the circle graph. Write the ratio of the dollar amounts as lowest-term fractions.

$\dfrac{10}{30} = \dfrac{1}{3}$ **7.** tithe to recreation

$\dfrac{15}{24} = \dfrac{5}{8}$ **8.** gas to clothing

$\dfrac{15}{17}$ **9.** gas to miscellaneous

$\dfrac{17}{96}$ **10.** miscellaneous to total monthly expenses

$\dfrac{24}{96} = \dfrac{1}{4}$ **11.** clothing to total monthly expenses

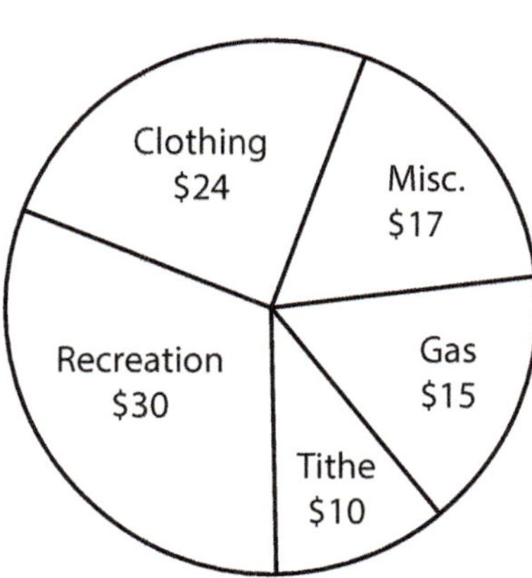

Last year the school soccer team won 12 games and lost 6 games. Write the ratio as a lowest-term fraction.

$\frac{6}{12} = \frac{1}{2}$ **12.** games lost to games won

$\frac{12}{18} = \frac{2}{3}$ **13.** games won to games played

$\frac{18}{6} = \frac{3}{1}$ **14.** games played to games lost

$\frac{12}{6} = \frac{2}{1}$ **15.** games won to games lost

Determine whether the given ratios are equivalent. If so, write the lowest term ratio.

no **16.** $\frac{5}{12}$ and $\frac{29}{72}$

yes; $\frac{2}{3}$ **17.** $\frac{12}{18} = \frac{56}{84}$

yes; $\frac{17}{4}$ **18.** $\frac{17}{4} = \frac{85}{20}$

no **19.** $\frac{3}{51} = \frac{9}{102}$

Complete each ratio table.

20.

meters	13	52	156
seconds	1	4	12

21.

dollars	8.25	49.50	74.25
burgers	1	6	9

22.

milliliters	237	474	1422
cups	1	2	6

23.

apples	13.6	34	81.6
pies	2	5	12

24.

batches	1.5	3	3.75
hours	2	4	5

25.

Yen	226	678	5424
USD	2	6	48

Unit Pricing

Name _______________________________

Use a calculator to find the unit price of each item. If necessary, round to the nearest tenth of a cent. Indicate the best buy with an asterisk.

Item		Size	Price	Unit Price	*Best Buy
Example: salad dressing		8 oz	$1.98	24.8¢ per oz	
		16 oz	$2.96	18.5¢ per oz	*
1.	detergent	50 oz	$2.87	5.7¢ per oz	*
		130 oz	$7.97	6.1¢ per oz	
2.	soft drink	2 L (67.6 oz bottle)	$1.25	1.8¢ per oz	*
		24 (12-oz) cans	$6.98	2.4¢ per oz	
3.	juice	64 oz	$2.48	3.9¢ per oz	
		1 gal (128 oz)	$3.98	3.1¢ per oz	*
4.	chewing gum	15 pieces	$1.00	6.7¢ per piece	
		3-pack (45 pieces)	$2.48	5.5¢ per piece	*
5.	cheese	16 oz	$3.58	22.4¢ per oz	
		5 lb (80 oz)	$15.47	19.3¢ per oz	*
6.	french fries	26 oz	$1.68	6.5¢ per oz	*
		32 oz	$2.58	8.1¢ per oz	
7.	crackers	21 oz	$4.54	21.6¢ per oz	*
		30 1-oz packs	$9.92	33.1¢ per oz	
8.	milk	$\frac{1}{2}$ gal	$1.55	2.4¢ per oz or 19.4¢ per cup	
		1 gal	$2.42	1.9¢ per oz or 15.1¢ per cup	*
9.	carrots	2 lb bag (32 oz)	$1.92	6¢ per oz	*
		2 lb bag (organic)	$2.86	8.9¢ per oz	
10.	peanut butter	18 oz (store brand)	$1.18	6.6¢ per oz	*
		28 oz (name brand)	$3.84	13.7¢ per oz	

11. Can you think of a couple of examples from this list when it would be better stewardship of your resources to buy the item with the higher unit price? Answers will vary. Examples: The larger item could go bad before you could use it all. The individual packs lead to less waste or portion control. One brand is just better to you or for you.

Cooking for a Crowd
(Mixed Practice; use after Section 7.2)

To raise money for a summer mission trip, the youth group mission team sold 60 tickets to a fundraiser brunch for $7 each. The items served for the fundraiser will be a breakfast casserole, doughnuts (1 per person) and orange juice (1 cup per person).

Fill in the blanks below to determine how much of each ingredient is needed for 60 guests.

Easy Breakfast Casserole

24 oz	frozen shredded hash browns
16 oz	cubed ham
8 oz	shredded cheese
12	eggs
1 c	milk

Salt and pepper to taste

Preheat oven to 350°. Spray 13 × 9 in. baking dish. Mix hash browns, ham, and cheese in a large bowl. Pour into baking dish. Whisk the eggs, milk, salt, and pepper in the mixing bowl. Pour over the hash brown mixture. Do not cover. Bake for 1 hour. Makes 12 servings.

1. 120 oz or $7\frac{1}{2}$ lb
2. 80 oz or 5 lb
3. 40 oz or $2\frac{1}{2}$ lb
4. 60 eggs
5. 5 c, 1.25 qt, $\frac{5}{16}$ gal, or 40 oz

Use the unit costs of the ingredients (including the doughnuts and orange juice) to calculate the cost of each line item.

Item	Unit cost		Answer	
hash browns	9¢ per oz	6.	$10.80	
cubed ham	42¢ per oz	7.	$33.60	
shredded cheese	17¢ per oz	8.	$6.80	
eggs	15¢ each	9.	$9.00	
milk	16¢ per cup	10.	$0.80	
orange juice	25¢ per cup	11.	$15.00	need 60 cups
doughnuts	33¢ each	12.	$19.80	need 60 doughnuts
total cost for ingredients		13.	$95.80	
cost for ingredients per person		14.	$1.60	

15. Determine the amount of money that will be earned through this fundraiser.

$324

Rates and Ratios

Name ______________________________

Use ratios or proportions to solve the following word problems.

1. Mrs. Burke has 27 students in her class. Nine of her students are left-handed. What is the ratio of left-handed students to right-handed students? $\frac{9}{18} = \frac{1}{2}$

2. A Boy Scout troop hiked 12 km in 3 hr. At this rate, how far can they hike in 5 hr? 20 km

3. At Oakridge Christian School, the ratio of students to teachers is 55 to 2. If there are 275 students in the school, how many teachers are there? 10 teachers

4. Last month Mrs. Hendericks purchased 500 sheets of paper for $3.25. At that rate, how much will 750 sheets of paper cost? $4.88

5. While traveling from Indianapolis to Chicago, Mr. Lawrence traveled 180 mi on 6 gal of gasoline. How many gallons will it take to travel the 330 mi from Chicago to Cleveland, Ohio? 11 gal

6. If Mr. Lawrence paid $8 for 6.4 gal of gasoline, how much would it cost him to buy 14.2 gal? $17.75

Proportion

(Enrichment; use after Section 7.3)

Part A Activity

Use sticks with the following lengths: 1 ft, 2 ft, and 3 ft. Measure and record in the table below the shadows created by the sticks. Measure the shadows of the taller objects listed in exercises 4–6 and write proportions to find the height of these taller objects.

	Height of Stick in Feet	Length of Shadow in Feet
1.	1	Answers will vary.
2.	2	
3.	3	

	Taller Object	Height of Taller Object in Feet	Length of Shadow in Feet
4.	flagpole	Answers will vary.	Answers will vary.
5.	school building		
6.	tree		

Part B Activity

Closing your left eye, hold a 6 in. rod exactly 12 in. from your right eye. Move until the rod completely covers the exact height of the object you are measuring. Measure the distance from the object to the eye. This gives the proportion used to find the height of the desired object. Study the example below. Find the height of the objects listed in the table. If necessary, substitute other objects for the given objects.

	Object	Distance from the Object to the Eye in Inches	Proportion	Height of the Object in Inches	Height of the Object in Feet
	Example	300 in.	$\frac{6}{12} = \frac{n}{300}$	150 in.	$12\frac{1}{2}$ ft
7.	flagpole	Answers will vary.			
8.	school building				
9.	tree				
10.	utility pole				

Solve for n in each proportion.

$n = 5$ **11.** $\frac{n}{8} = \frac{30}{48}$

$n = 12$ **12.** $\frac{10}{24} = \frac{5}{n}$

$n = 10$ **13.** $\frac{5}{n} = \frac{8}{16}$

$n = 15$ **14.** $\frac{12}{20} = \frac{9}{n}$

Write a proportion and solve it.

$\frac{1}{6} = \frac{n}{18}; n = 3$ **15.** 1 is to 6 as n is to 18.

$\frac{4}{n} = \frac{24}{42}; n = 7$ **16.** 4 is to n as 24 is to 42.

Setting Up to Scale

(Enrichment; use after Section 7.4)

Name ________________________________

The youth group is planning carnival games for the church's
Fall Family Festival in October. The area designated for the carnival
is a 48 × 30 ft space. They have drawn the following 6 × 3.75 in. rectangle
to represent the carnival area:

Determine the scale factor used for the drawing, then find the dimensions on the drawing for each carnival item. 1 in.: 8 ft

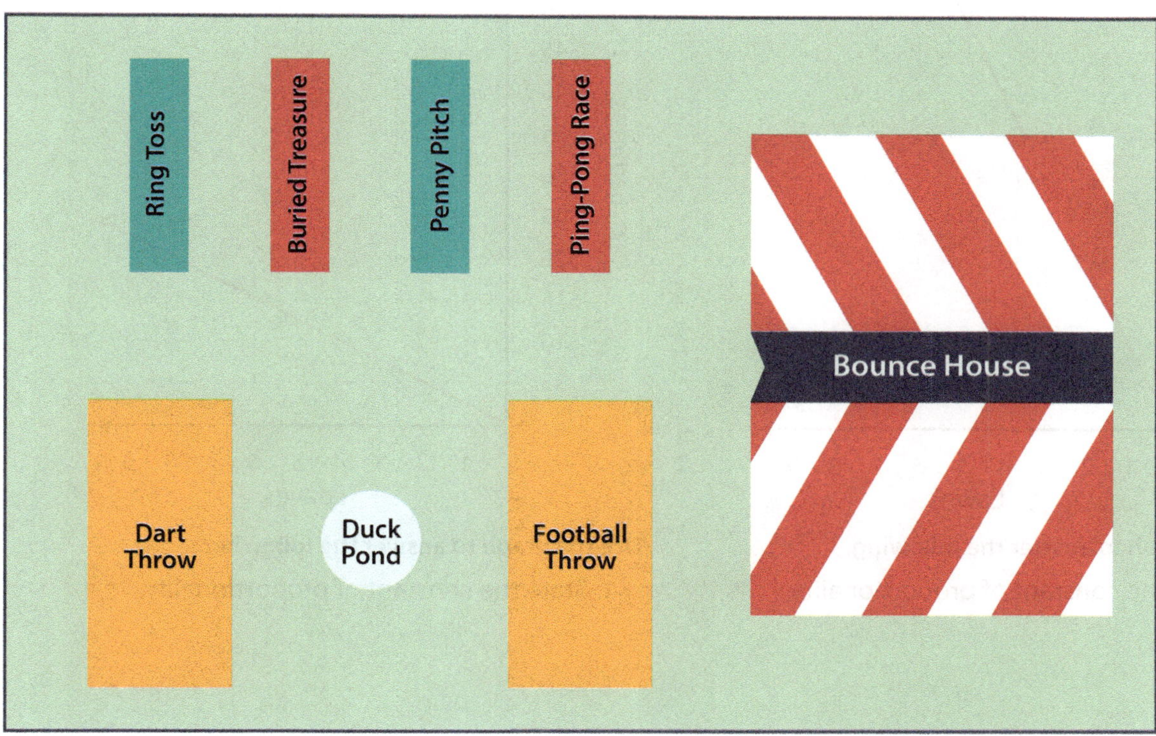

$1\frac{7}{8} \times 2\frac{1}{2}$ in.
1. a 15 × 20 ft bounce house

$1 \times \frac{5}{16}$ in. each
2. 4 folding game tables that are 8 × 2.5 ft each

$1\frac{1}{2} \times \frac{3}{4}$ in.
3. 2 larger game areas for Ring Toss, Buried Treasure, Penny Pitch, and Ping-Pong Race. Each area is 12 × 6 ft.

$\frac{1}{2}$ in. diameter
4. a kiddie pool with a diameter of 4 ft

Cut out shapes for each activity and determine the best arrangement for the event space.

Graphing Potential Earnings
(Mixed Practice; use after Section 7.5)

Graph Sean's potential earnings if he charges $10 for each lawn that he mows.

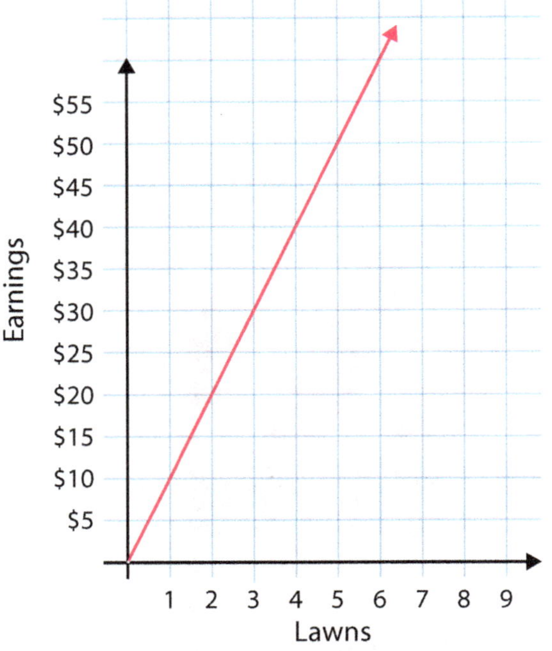

Graph Lexi's potential earnings if she sells 2 loaves of bread for $5.

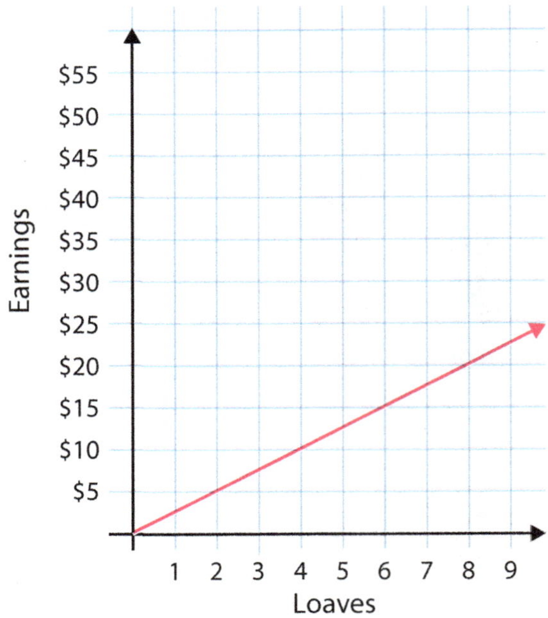

Use the graph to answer the following.

1. State the constant of proportionality. $k = 10$

2. How many lawns does Sean need to mow to earn $50? 5 lawns

3. How much money will Sean earn if he mows 12 lawns? $120

Use the graph to answer the following.

4. State the constant of proportionality. $k = 2.5$

5. How many loaves does Lexi need to bake to earn $20? 8 loaves

6. How much money will Lexi earn if she bakes 24 loaves? $60

Fundamentals of Math

Chapter 7 Cumulative Review

Name _______________________________

_____C_____ 1. What is the value of the underlined digit in 5<u>8</u>1,426? **[1.1]**

      A. 800,000   **B.** 8000

      C. 80,000   **D.** not given

_____C_____ 2. Estimate the difference: $723 - 285$. **[1.2]**

      A. 300   **B.** 420

      C. 400   **D.** not given

_____D_____ 3. Express 3^4 in standard form. **[1.5]**

      A. 12   **B.** 64

      C. 27   **D.** not given

_____A_____ 4. Simplify $25 + (-17) + (-2)$. **[1.7]**

      A. 6   **B.** 44

      C. −10   **D.** not given

_____B_____ 5. Simplify $3(-9)$. **[2.5]**

      A. −6   **B.** −27

      C. 27   **D.** not given

_____D_____ 6. Simplify $\sqrt{3^2 + 4^2} + (-2)^3$. **[2.7]**

      A. 13   **B.** 17

      C. 33   **D.** not given

_____B_____ 7. Name the property illustrated by $45 \times 1 = 45$. **[3.2]**

      A. Associative   **B.** Identity

      C. Commutative   **D.** not given

_____C_____ 8. Simplify the following expression: $5a + 2b - 3a - 4b + 6$. **[3.5]**

      A. 6   **B.** $4ab + 6$

      C. $2a - 2b + 6$ **D.** not given

_____C_____ 9. Give the answer in scientific notation. $(3.45 \times 10^6 + 2.71 \times 10^6) \times (6.2 \times 10^7)$ **[3.8]**

      A. 3.82×10^{13} **B.** 3.82×10^{12}

      C. 3.82×10^{14} **D.** not given

_____D_____ 10. The number 18,252 is divisible by which of the following primes? **[4.1]**

      A. 2, 3, 17   **B.** 2, 3, 19

      C. 2, 3, 11   **D.** 2, 3, 13

______B______**11.** Find the greatest common factor of 112 and 84. **[4.2]**

    A. 336   **B.** 28

    C. 7   **D.** not given

______A______**12.** Find the decimal equivalent of $\frac{11}{16}$. [4.5]

    A. 0.6875   **B.** 0.6957

    C. 0.6842   **D.** not given

______C______**13.** Add $\frac{3}{8} + \frac{3}{4} + \frac{1}{3}$. **[5.1]**

    A. $\frac{7}{24}$ **B.** $1\frac{7}{15}$

    C. $1\frac{11}{24}$ **D.** not given

______B______**14.** Subtract $12 - 9\frac{4}{5}$. **[5.3]**

    A. $3\frac{4}{5}$ **B.** $2\frac{1}{5}$

    C. $2\frac{4}{5}$ **D.** not given

______B______**15.** Identify which of the following numbers is irrational. **[5.6]**

    A. $\sqrt{\frac{100}{4}}$ **B.** $\sqrt{\frac{49}{9}}$

    C. $\sqrt{\frac{48}{3}}$ **D.** all are rational

______C______**16.** Which of the expressions below is equivalent to $\frac{2}{3}(15a + 6)$? **[5.7]**

    A. $14a$ **B.** $30a + 12$

    C. $10a + 4$ **D.** not given

______D______**17.** Solve for x: $18 - x = 3$. **[6.2]**

    A. $x = -15$ **B.** $x = 21$

    C. $x = -21$ **D.** not given

______B______**18.** Solve for y: $2(3y - 4) = 16$. **[6.5]**

    A. $y = 2$ **B.** $y = 4$

    C. $y = 8$ **D.** not given

______C______**19.** Solve for w: $-3w > 15$. **[6.7]**

    A. $w > -5$ **B.** $w \geq -5$

    C. $w < -5$ **D.** $w \leq -5$

______B______**20.** Solve for a: $\frac{a}{5} - 6 > -26$. **[6.8]**

    A. $a > -4$ **B.** $a > -100$

    C. $a < -4$ **D.** $a < -100$

Renaming Percents
(Extra Practice; use after Section 8.1)

Name _______________________________

Write each number as a percent. For exercises 1–12, express parts of a percent as decimals. For exercises 13–21, express them as fractions.

___2%___ **1.** 0.02

___125%___ **2.** 1.25

___36%___ **3.** 0.36

___30%___ **4.** 0.3

___250%___ **5.** 2.5

___32.4%___ **6.** 0.324

___101%___ **7.** 1.01

___62.5%___ **8.** 0.625

___300%___ **9.** 3

___3.4%___ **10.** 0.034

___4160%___ **11.** 41.6

___0.5%___ **12.** 0.005

___60%___ **13.** $\frac{3}{5}$

___$26\frac{2}{3}\%$___ **14.** $\frac{4}{15}$

___85%___ **15.** $\frac{17}{20}$

___$55\frac{5}{9}\%$___ **16.** $\frac{5}{9}$

___$2\frac{1}{2}\%$___ **17.** $\frac{1}{40}$

___175%___ **18.** $\frac{7}{4}$

___28%___ **19.** $\frac{7}{25}$

___$90\frac{10}{11}\%$___ **20.** $\frac{10}{11}$

___15%___ **21.** $\frac{3}{20}$

For exercises 22–30, change the percent to both a decimal and a fraction. If necessary, use a repeating bar to indicate repeating decimals.

	Percent	Decimal	Fraction
22.	36%	0.36	$\frac{9}{25}$
23.	15%	0.15	$\frac{3}{20}$
24.	250%	2.5	$\frac{5}{2}$
25.	$6\frac{2}{3}\%$	$0.0\overline{6}$	$\frac{1}{15}$
26.	0.25%	0.0025	$\frac{1}{400}$
27.	90%	0.9	$\frac{9}{10}$
28.	$33\frac{1}{3}\%$	$0.\overline{3}$	$\frac{1}{3}$
29.	4%	0.04	$\frac{1}{25}$
30.	135%	1.35	$\frac{27}{20}$

Sales Tax

Name _______________________________

Complete the following restaurant checks by filling in any missing amounts. The sales tax rate is 5%. Round to the nearest cent if necessary.

1.

1 fish sandwich	$1.29
1 coleslaw	$0.75
1 milk	$0.60
subtotal	$2.64
tax	$0.13
total	$2.77

2.

1 steak dinner	$11.50
1 spaghetti dinner	$8.75
2 desserts @ $3.25 each	$6.50
2 drinks @ $1.10 each	$2.20
subtotal	$28.95
tax	$1.45
total	$30.40

When finding the total cost, you pay 100% for the item plus the rate of the sales tax. If the sales tax rate is 5%, you actually pay 105% of the original cost.

To find the total cost of a $21.95 shirt with a 5% tax rate, multiply $21.95 by 105%.

$$21.95 \times 105\% = 21.95 \times 1.05 = 23.0475$$

The total cost (rounded to the nearest cent) is $23.05.

Find the total cost rounded to the nearest cent.

3. clock radio: $24.75
sales tax rate: 6%
$26.24

4. bicycle: $415.00
sales tax rate: 4.5%
$433.68

5. eraser: $0.79
sales tax rate: 5%
$0.83

6. used car: $4900
sales tax rate: 5%
$5145

Finding Percents

Use the data in the table to answer the following questions. If necessary, round each percent to the nearest whole percent.

Baseball Statistics				
Players	At Bats	Hits	Home Runs	Strikeouts
Daniel	70	22	4	12
Mark	68	24	3	11
Chris	75	28	4	14
Eric	40	13	3	8

______35%______ **1.** What percent of Mark's at bats were hits?

______5%______ **2.** What percent of Chris's at bats were home runs?

______17%______ **3.** What percent of Daniel's at bats were strikeouts?

______Eric______ **4.** Who had the higher percentage of home runs, Eric or Daniel?

______Chris______ **5.** Who had the higher percentage of hits, Mark or Chris?

______Chris______ **6.** Who had the lower percentage of strikeouts, Chris or Eric?

______Chris______ **7.** Who had the highest percentage of hits?

______Eric______ **8.** Who had the highest percentage of home runs?

______Eric______ **9.** Who had the highest percentage of strikeouts?

______Mark______ **10.** Who had the lowest percentage of strikeouts?

Percents in Action—The Cost of Pet Ownership

Name ___________________________

(Mixed Practice; use after Section 8.3)

Troy asked his dad if he could get a puppy. His dad asked him to research how much it would cost to own a medium-sized dog and what he would need to do to care for it. After a lot of research on the Internet, he found that dogs live about 10 years and that the first year costs more than the remaining 9 years.

Use the circle graph below to answer Troy's questions.

A. training
B. vaccinations/shots
C. veterinarian fees
D. fleas/ticks/worm treatments
E. equipment like doghouse, collar, leash
F. food and chew toys

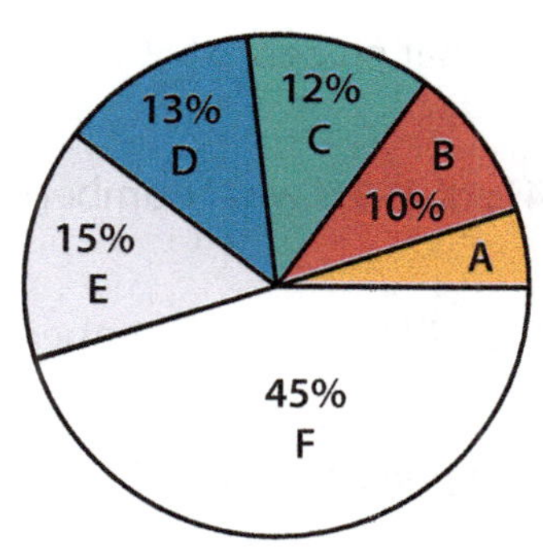

___$700___ **1.** Food and chew toys cost $315/yr. What is the average yearly cost to own a dog?

___$7000___ **2.** What is the 10 yr cost to own a dog?

___5%___ **3.** Even though training may take place early in the puppy's life, what percent of the average yearly cost is this expense?

___$105___ **4.** What is the average yearly cost for equipment?

___$245___ **5.** What is the average yearly cost for all health expenses (veterinarian, vaccinations, and flea/tick/worm treatments)?

___$58.34___ **6.** If Troy is required to pay for all of the dog's expenses, how much will he have to budget per month from his allowance and part-time job?

7. When Troy showed his dad the costs, his dad reminded him that his puppy would have to be cared for each summer while they were on vacation. After more checking, Troy found that a kennel would cost $180 per year. Identify this new category as *G*, and find the new percent for each category. Round to the nearest whole percent.   A) 4%; B) 8%; C) 10%; D) 10%; E) 12%; F) 36%; G) 20%

___$73.34___ **8.** Including kennel costs, how much will Troy need to budget per month to take care of his puppy's needs?

___42%___ **9.** Troy mows three lawns each week at $20 per lawn. His mowing jobs usually last for about 30 weeks out of the year. His allowance is $25 per month. To the nearest percent, what part of his yearly income will be used to care for the dog?

Percent Equations

Solve.

______54______ **1.** 18% of 300 is what number?

______25______ **2.** 4 is 16% of what number?

______14%______ **3.** 7 is what percent of 50?

______35______ **4.** 14% of 250 is what number?

______200______ **5.** 25 is 12.5% of what number?

______22%______ **6.** What percent of 50 is 11?

______75______ **7.** 21 is 28% of what number?

______76______ **8.** What is 95% of 80?

______68%______ **9.** 119 is what percent of 175?

______210______ **10.** 40% of what number is 84?

______30%______ **11.** What percent of 50 is 15?

______45.84______ **12.** Find 12% of 382.

______650______ **13.** 22% of what number is 143?

______405______ **14.** 54% of 750 equals what?

______50______ **15.** 70% of n is 35.

______38%______ **16.** 24.7 is p% of 65.

______81.6______ **17.** What is 68% of 120?

______1250______ **18.** 48% of what number is 600?

Simple Interest

(Mixed Practice; use after Section 8.4)

To calculate simple interest use the formula $I = Prt$. I stands for the amount of interest received, P for the amount invested or borrowed, r for the rate (percent) of interest per year, and t for the time (in years). The amount in the account, A, is the principal plus the amount of interest accrued.

Complete the table. Round to the nearest cent.

	Principal	Rate	Time	Interest	Amount
1.	$3800	7%	2 years	$532	$4332
2.	$9000	12%	3 years	$3240	$12,240
3.	$12,000	5%	6 months	$300	$12,300
4.	$20,000	$9\frac{1}{2}$%	4 years	$7600	$27,600
5.	$700	6.2%	18 months	$65.10	$765.10
6.	$6500	4.75%	30 months	$771.88	$7271.88

Complete the table. The interest is left in the account when it is paid each year, so the amount in the account increases each year. Round the interest to the nearest cent.

	6% Annual Interest	Amount Owed on the Account	Interest for Year
7.	year 1	$500	$30
8.	year 2	$530	$31.80
9.	year 3	$561.80	$33.71
10.	year 4	$595.51	$35.73
11.	year 5	$631.24	$37.87
12.	year 6	$669.11	$40.15

_$709.26_ **13.** What is the total amount in the account after 6 years?

_$209.26_ **14.** What is the total amount of interest earned after 6 years?

Photo Sizing

<u>reduction</u> **1.** Does 72% give an enlargement or a reduction?

Give the correct copier settings (percents) for the following.

<u>175%</u> **2.** $1\frac{3}{4}$

<u>40%</u> **3.** $\frac{2}{5}$

<u>67%</u> **4.** $\frac{2}{3}$

<u>200%</u> **5.** 2

<u>50%</u> **6.** Reduce a 4 × 6 in. photograph to make the larger dimension 3 in.

<u>2 in.</u> **7.** Find the smaller dimension of the photo in question 6.

<u>200%</u> **8.** Enlarge a 4 × 6 in. photograph to make the smaller dimension 8 in.

<u>12 in.</u> **9.** Find the longer dimension of the photo in question 8.

<u>$16\frac{2}{3}$%</u> **10.** A typical photo size is 8 × 10. If you are enlarging a 4 × 6 photo to 8 × 10, what percent of the longer dimension will you need to crop out?

Percent Change

Name ___________________________

Find the percent of change and specify increase or decrease.

60% decrease
_________________ **1.** original: 300; new amount: 120

15% increase
_________________ **2.** original: 140; new amount: 161

28% increase
_________________ **3.** from 75 to 96

2% decrease
_________________ **4.** from 150 to 147

6% increase
_________________ **5.** from \$8.00 to \$8.48

34% decrease
_________________ **6.** from 600 to 396

135% increase
_________________ **7.** from 120 to 282

15% decrease
_________________ **8.** from \$19.60 to \$16.66

Find the percent change. Specify as increase or decrease.

9. The temperature at midnight was 28°F. If the temperature was 49°F at noon, what was the percent of change? 75% increase

10. Mr. Clark's electric bill was \$125 in August. If the bill was \$112.50 in September, what was the percent of change? 10% decrease

Weather Changes

(Enrichment; use after Section 8.6)

Name ____

Find a website that shows weather history in your area. You can enter your zip code and find current conditions, as well as a history tab that allows you to look back at any date for which there are records.

__Answers will vary.__   **1.** What is the current temperature in your city?

__Answers will vary.__   **2.** What is the high temperature forecasted for tomorrow?

__Answers will vary.__   **3.** What is the record high for your city on today's date?

__Answers will vary.__   **4.** What was the high temperature 1 year ago today?

__Answers will vary.__   **5.** What was the high temperature 6 months ago today?

__Answers will vary.__   **6.** What was the high temperature on the day you were born? (If you were born in a different city than where you live now, be sure to look up the correct city.)

Calculate the percent changes indicated below. Be sure to indicate if they represent an increase or a decrease:

__Answers will vary.__   **7.** from the current temperature to tomorrow's high

__Answers will vary.__   **8.** from the current temperature to the record high for this date

__Answers will vary.__   **9.** from the high temperature 1 year ago today to tomorrow's high

  10. from the high temperature 6 months ago today to tomorrow's high

  11. from the high temperature on your birthday to tomorrow's high

Fundamentals of Math

Retail Sales

Name ______________________________

In September, Cora received a new shipment of winter outerwear for her clothing shop. Her first job was to print the price tags before she could put them out on the racks. Her store does a 60% markup on their cost for goods.

Fill in the table with the retail prices based on a 60% markup.

		Cost	Retail Price
1.	coats	$49	$78.40
2.	boots	$15	$24.00
3.	gloves	$4.50	$7.20

In February, Cora did an analysis of sales of the winter outerwear from the month of January.

Fill in the table with the percent changes for each item. Round to the nearest whole percent. Indicate whether it represents an increase or a decrease.

		Percent Change	Increase/Decrease
4.	Sales of coats went from 67 to 42 coats.	37%	Decrease
5.	Sales of boots went from 52 to 34 pairs.	34%	Decrease
6.	Sales of gloves went from 107 to 120 pairs.	12%	Increase

Cora decided to have a 20% off sale over Presidents' Day weekend to move as much of the winter stock as possible. Janet bought a coat, two pairs of gloves, and a pair of boots. She used a coupon for 10% off one item.

Fill in Janet's sales receipt.

	Item	Count	Retail Price	Discount Price
7.	coat	1	$78.40	$56.45
8.	gloves	2	$14.40	$11.52
9.	boots	1	$24.00	$19.20
10.	Total			$87.17

___yes___ **11.** Even with all of the discounts, is Janet still paying more than the store's original cost for the items?

Chapter 8 Cumulative Review

Name ________________________

____B____ **1.** Estimate the product of 418 × 36. **[1.3]**

    A. 12,000   **B.** 16,000

    C. 20,000   **D.** not given

____C____ **2.** Simplify $-5 - (-5)$. **[2.4]**

    A. −10   **B.** 10

    C. 0   **D.** not given

____A____ **3.** Simplify $(-100 \div 5) \div (-2)$. **[2.6]**

    A. 10   **B.** −10

    C. 40   **D.** not given

____C____ **4.** Evaluate $14 - 5x$ if $x = 3$. **[3.1]**

    A. 27   **B.** 1

    C. −1   **D.** not given

____B____ **5.** Find the missing number in $4 \times (8 + 17) = (4 \times \underline{}) + (4 \times 17)$. **[3.3]**

    A. 4   **B.** 8

    C. 17   **D.** not given

____C____ **6.** Simplify the following expression: $(4n - 2) + (5n - 8)$. **[3.6]**

    A. $9n - 6$ **B.** $9n + 6$

    C. $9n - 10$ **D.** not given

____B____ **7.** Find the least common multiple of 26 and 169. **[4.3]**

    A. 13   **B.** 338

    C. 4394   **D.** not given

____C____ **8.** Which fraction is not equivalent to the others? $\frac{3}{7}, \frac{6}{14}, \frac{12}{28}, \frac{18}{42}, \frac{45}{98}$ **[4.4]**

    A. $\frac{12}{28}$ **B.** $\frac{18}{42}$

    C. $\frac{45}{98}$ **D.** All are equivalent.

____D____ **9.** Order the following fractions from least to greatest: $\frac{3}{8}, \frac{5}{16}, \frac{9}{32}$. **[4.6]**

    A. $\frac{3}{8}, \frac{9}{32}, \frac{5}{16}$ **B.** $\frac{5}{16}, \frac{3}{8}, \frac{9}{32}$

    C. $\frac{9}{32}, \frac{3}{8}, \frac{5}{16}$ **D.** not given

____A____ **10.** Subtract $\frac{17}{20} - \frac{5}{12} - \frac{2}{15}$. **[5.2]**

    A. $\frac{3}{10}$ **B.** $\frac{9}{60}$

    C. $\frac{11}{15}$ **D.** not given

_____**B**_____**11.** Find the product of $\frac{12}{15} \times \frac{5}{18}$. Express the answer in lowest terms. **[5.4]**

    A. $\frac{60}{270}$ **B.** $\frac{2}{9}$

    C. $\frac{25}{72}$ **D.** not given

_____**A**_____**12.** Simplify $\left(\frac{7}{8} - \frac{5}{9}\right) \cdot 6$ to a reduced rational number. **[5.6]**

    A. $\frac{23}{12}$ **B.** $\frac{138}{12}$

    C. $\frac{3}{1}$ **D.** not given

_____**B**_____**13.** Translate the following sentence into an algebraic equation: "Five less than a number is seventeen." **[6.1]**

    A. $5 - n = 17$ **B.** $n - 5 = 17$

    C. $17 - 5 = n$ **D.** not given

_____**C**_____**14.** Solve for w: $\frac{w}{5} = 1.3$. **[6.3]**

    A. 0.26   **B.** 3.85

    C. 6.5   **D.** not given

_____**A**_____**15.** Which of the following numbers satisfies this inequality: $x - 2 > 7$? **[6.6]**

    A. 10   **B.** 9

    C. 8   **D.** choices A and B

_____**D**_____**16.** The school orchestra has 12 seventh graders and 15 eighth graders. Write the ratio of seventh graders to eighth graders as a ratio in reduced form. **[7.1]**

    A. $5:4$ **B.** $5:9$

    C. $4:9$ **D.** not given

_____**A**_____**17.** Simplify: $\dfrac{\frac{1}{4}}{\frac{2}{3}}$. **[7.2]**

    A. $\frac{3}{8}$ **B.** $\frac{1}{6}$

    C. $2\frac{2}{3}$ **D.** not given

_____**B**_____**18.** The Statue of Liberty measures 305 ft from the foundation to the tip of her torch. The tablet she is holding is 23.6 ft long. Amanda bought a souvenir replica of the statue that is 8 in. tall. How long is this statue's tiny tablet? Round to the nearest hundredth of a foot. **[7.4]**

    A. 0.55 in.   **B.** 0.62 in.

    C. 1.62 in.   **D.** not given

_____**B**_____**19.** Determine the constant of proportionality in the following ratio: $15:10$. **[7.5]**

    A. $k = 5$ **B.** $k = \frac{2}{3}$

    C. $k = \frac{3}{2}$ **D.** not given

9 Measurement

Customary Unit Conversions
(Extra Practice; use after Section 9.2)

Name ___________________________________

Complete each conversion.

1. 3 ft = _____36_____ in.

2. 2 gal = _____32_____ c

3. 2 mi = _____3520_____ yd

4. 6 c = _____$1\frac{1}{2}$_____ qt

5. 12 ft = _____4_____ yd

6. 6 qt = _____$1\frac{1}{2}$_____ gal

7. 60 in. = _____5_____ ft

8. 2 lb = _____32_____ oz

9. 21,120 ft = _____4_____ mi

10. 6000 lb = _____3_____ tn

11. 3 mi = _____15,840_____ ft

12. 64 oz = _____4_____ lb

13. 1 yd = _____36_____ in.

14. 3 mi = _____190,080_____ in.

15. 48 in. = _____$1\frac{1}{3}$_____ yd

16. 1 tn = _____32,000_____ oz

17. 30 in. = _____$2\frac{1}{2}$_____ ft

18. 1 gal = _____128_____ fl oz

19. 2 c = _____16_____ fl oz

20. 88 oz = _____$5\frac{1}{2}$_____ lb

Measurements of Bible Times

Look up the following Bible references to find the missing dimensions. Change the Bible-time measurement to a modern equivalent as instructed in each question.

The cubit was a measurement used during Bible times based on the distance from the end of the elbow to the tip of the middle finger. On this worksheet, use an average measurement of 18 in. for the cubit.

1. In Genesis 6:15, the dimensions of Noah's ark are given. The ark was __________ cubits long, __________ cubits wide, and __________ cubits high. Find the dimensions in feet.
   300 cubits = 450 ft; 50 cubits = 75 ft; 30 cubits = 45 ft

2. In Exodus 37:1, the builders of the ark of the covenant were instructed to build it __________ cubits long, __________ cubits wide, and __________ cubits high. Find the dimensions in feet. $2\frac{1}{2}$ cubits = $3\frac{3}{4}$ ft; $1\frac{1}{2}$ cubits = $2\frac{1}{4}$ ft; $1\frac{1}{2}$ cubits = $2\frac{1}{4}$ ft

The furlong is a measurement used for longer distances. A furlong is equal to approximately 400 cubits, or 600 ft.

3. Revelation 21:16 reveals the dimension of the New Jerusalem. The city is described as a cube with each side being how many furlongs? Change this dimension to feet and then to miles (to the nearest tenth of a mile). 12,000 ≈ 7,200,000 ft ≈ 1363.6 mi

The firkin is a measure of capacity. A firkin is equivalent to approximately 9 gallons.

4. In John 2:6, we read that at the wedding at Cana "there were set there six waterpots of stone … containing two or three firkins apiece." Change 2 firkins to gallons, and then to cups. 2 firkins ≈ 18 gal = 288 c

Other measures of capacity are the ephah (sometimes just called a "measure" or a "deal"), which was equal to approximately 6 gal (3/5 of a bushel) and the hin (about the same as a gallon).

5. In Exodus 29:38–40, the daily sacrifice is instructed to be "two lambs of the first year day by day continually. The one lamb thou shalt offer in the morning; and the other lamb thou shalt offer at even: And with the one lamb a tenth deal [tenth of an ephah] of flour mingled with the fourth part of an hin of beaten oil; and the fourth part of an hin of wine for a drink offering." How many cups of flour, oil, and wine would have been required every day for the sacrifice (to the nearest tenth of a cup)?
   19.2 cups of flour. 8 cups each of oil and wine.

Metric Units of Length

Name ______________________________

Measure each segment below with a metric ruler. Record the length to the nearest centimeter and the nearest millimeter.

0.9 cm; 9 mm _______________ **1.** ____

3 cm; 30 mm _______________ **2.** _______________

1.8 cm; 18 mm _______________ **3.** _________

6 cm; 60 mm _______________ **4.** ___________________________

5.4 cm; 54 mm _______________ **5.** ____________________________

Measure the objects below using a metric ruler or a meter stick.
Be careful to use the correct units. Answers will vary. You may want to provide items to ensure that answers are consistent.

_______________________ **6.** length of an unsharpened pencil

_______________________ **7.** width of your desk

_______________________ **8.** height of your desk

_______________________ **9.** length of your math book

_______________________ **10.** length of a piece of notebook paper

_______________________ **11.** width of classroom door

_______________________ **12.** length of a whiteboard

_______________________ **13.** length of the classroom's front wall

_______________________ **14.** width of the classroom floor

_______________________ **15.** your height

Renaming Metric Units

Even though some of the metric prefixes have fallen out of common usage, understanding that every power of 10 is represented can help make conversions within the metric system simple.

Example 1: Convert 43 km to centimeters.

Answer: On the staircase, you move 5 steps to go from kilometers to centimeters. Since you are moving down the staircase, changing from a large unit to a smaller unit, you must multiply (because centimeter is a smaller unit, more are needed). Each step represents a power of 10. Therefore, you multiply by 10^5.

$$43 \times 10^5 = 43 \times 100,000 = 4,300,000$$

Therefore, 43 km = 4,300,000 cm.

Remember that multiplying by a power of 10 can be done by moving the decimal point to the right. The number of places is determined by the power.

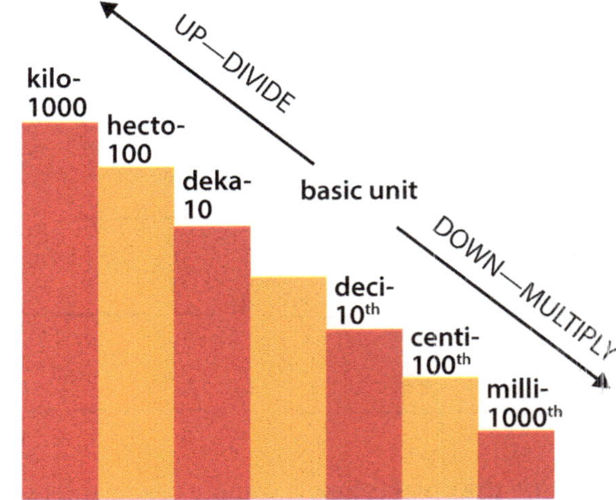

Example 2: Convert 490 cL to hectoliters.

Answer: On the staircase, to move from centiliter to hectoliter you must go up 4 steps. Moving up the staircase, changing from a smaller to a larger unit, requires you to divide (because hectoliter is a larger unit, fewer are needed). Since you have moved up 4 steps, divide by 10^4.

$$490 \div 10^4 = 490 \div 10,000 = 0.049$$

Therefore, 490 cL = 0.049 hL.

Remember that dividing by a power of 10 moves the decimal point to the left. The number of places is determined by the power.

Complete each conversion.

1. 4 m = __________ 400 __________ cm

2. 2.6 kg = __________ 260,000 __________ cg

3. 0.7 kL = __________ 700,000 __________ mL

4. 7400 mL = __________ 7.4 __________ L

5. 7 L = __________ 700 __________ cL

6. 6.15 mg = __________ 0.615 __________ cg

7. 2.06 km = __________ 2060 __________ m

8. 248 cg = __________ 2.48 __________ g

9. 14 kg = __________ 14,000 __________ g

10. 46.8 cm = __________ 468 __________ mm

11. 72 m = __________ 72,000 __________ mm

12. 142.68 mm = __________ 0.00014268 __________ km

Mass

(Enrichment; use after Section 9.3)

There are 3 things about the way the metric system was developed that were designed to make it easy to learn and use:

- It started with a unit for length that was based on an outside standard (distance from the equator to the North Pole divided by 10,000,000) instead of the traditional lengths of body parts or how far someone could walk in a day.

- Smaller and larger units are related by powers of 10, which makes converting easier.

- The systems for length, mass, and volume are interrelated. A cubic centimeter holds one milliliter, and a milliliter of water weighs 1 gram.

The following activity uses the relationship between milliliters and grams to weigh objects without a scale.

When a floating object is placed into water, it displaces (pushes out of the way) 1 mL of water for every 1 g of mass. Using a graduated cylinder and water, you can find the mass of a floating object by following these steps:

1. Fill the cylinder to a convenient level like 100 mL.

2. Place an object into the cylinder.

3. Read the new milliliter level.

4. Subtract the original milliliter level from the new milliliter level.

Teacher note: This activity will require you to gather needed objects before class presentation. Feel free to substitute other items if these are not readily available.

The difference is the mass in grams. Remember 1 mL of water equals 1 g of mass.

If the object does not float, you need to make a "boat" out of something like a plastic cup. (If it sinks, you are measuring its *volume*, not its weight.) Be sure your cup is included in the original milliliter level.

Sometimes it is easier to see if you put in several of the same object (like 5 marbles), then divide your number of grams by the number of objects that you used.

Find the mass of the following objects. Answers will vary.

______________ 1. paper clip   ______________ 2. marble

______________ 3. ice cube   ______________ 4. wooden pencil

______________ 5. lip balm   ______________ 6. quarter

______________ 7. penny   ______________ 8. medicine container

______________ 9. rubber duck   ______________ 10. wooden block

What else might you like to "weigh" using this method?

Teacher note: The same activity could be repeated using sinking objects to find the volume.

Metric/Customary Unit Conversions

Complete each conversion. If needed, round each answer to the nearest tenth.

Approximate Customary and Metric Conversions		
Length	**Capacity**	**Weight or Mass**
1 in. ≈ 2.54 cm	1 fl oz ≈ 30 mL	1 oz ≈ 28.35 g
1 ft ≈ 30.48 cm	1 L ≈ 1.06 qt	1 kg ≈ 2.2 lb
1 mi ≈ 1.61 km	1 tsp ≈ 4.93 mL	1 mT ≈ 1.10 tn
1 m ≈ 39.37 in.		

1. 10 cm = _____3.9_____ in.

2. 200 km = _____124.2_____ mi

3. 15 mL = _____3.0_____ tsp

4. 60 qt = _____56.6_____ L

5. 42 kg = _____92.4_____ lb

6. 3 oz = _____85.1_____ g

7. 60 tn = _____54.5_____ mT

8. 15 m = _____49.2_____ ft

9. 6 yd = _____5.5_____ m

10. 4 in. = _____101.6_____ mm

11. 16 L = _____4.2_____ gal

12. 1120 mL = _____37.3_____ fl oz

13. 3 tn = _____2727.3_____ kg

14. 3 kg = _____105.6_____ oz

Converting Temperature

Name ___________________________

You can evaluate an algebraic expression to convert a Fahrenheit temperature to a Celsius temperature. Celsius temperature is represented by the expression $\frac{5}{9}(F - 32)$, where F represents the Fahrenheit temperature.

Example: Find the Celsius temperature equivalent to 41°F.

Answer:

$C = \frac{5}{9}(F - 32)$

$C = \frac{5}{9}(41 - 32)$ Substitute 41° for F.

$C = \frac{5}{9} \cdot 9$

$C = 5$

Therefore, 41°F = 5°C.

Find the Celsius temperatures. Use $C = \frac{5}{9}(F - 32)$.
If necessary, round the answer to the nearest degree.

$C = 15°$ **1.** $F = 59°$

$C = 20°$ **2.** $F = 68°$

$C = 35°$ **3.** $F = 95°$

$C = 12°$ **4.** $F = 53°$

You can also calculate the Fahrenheit temperature if you know the Celsius by using the expression $F = \frac{9}{5}C + 32$.

Find the Fahrenheit temperatures. Use $F = \frac{9}{5}C + 32$.
If necessary, round the answer to the nearest degree.

$F = 113°$ **5.** $C = 45°$

$F \approx 61°$ **6.** $C = 16°$

$F = 41°$ **7.** $C = 5°$

$F \approx 48°$ **8.** $C = 9°$

Vacation to Canada

Customary to Metric Conversion
1 mile ≈ 1.6 kilometers
1 gallon ≈ 3.8 liters
1 kilogram ≈ 2.2 pounds

- To rename smaller units as larger units *divide*.
- To rename larger units as smaller units *multiply*.

Solve. Use the conversion facts listed above and a calculator.

1. The Campbell family saw a sign that said the distance to Toronto was 160 km. How many miles is it to Toronto? (Round to the nearest mile.) 99 mi

2. It took 53 L of gasoline to fill the tank in the Campbells' car. How many gallons were necessary to fill the tank? (Round to the nearest gallon.) 14 gal

3. The Campbells traveled from Detroit to North Bay, Ontario. They averaged 55 mph for 8 hr. How many kilometers did they travel? (Round to the nearest kilometer.) 704 km

4. Mrs. Campbell went to the grocery store to buy hamburger for lunch. The package said it weighed 0.68 kg. How many pounds did it weigh? (Round to the nearest tenth of a pound.) 1.5 lb

5. Mrs. Campbell's recipe calls for 2 lb of cheese. How many kilograms should she buy? (Round to the nearest tenth.) 0.9 kg

Fundamentals of Math

Metric Recipe Conversion

Name _______________________________

Sylvia is using two recipes where the ingredients are given in metric units. Change the ingredients to customary measuring units that she has in her kitchen—teaspoons and cups ($\frac{1}{4}$, $\frac{1}{3}$, $\frac{1}{2}$, $\frac{2}{3}$, $\frac{3}{4}$, and 1). Round answers to the nearest measuring cup she can use.

	Metric Chocolate Chip Cookies		
	Metric amount	**Customary amount**	**Ingredient**
1.	550 mL	$2\frac{1}{3}$ c	flour
2.	5 mL	1 tsp	baking soda
3.	5 mL	1 tsp	salt
4.	250 mL	1 c	butter or margarine, softened
5.	175 mL	$\frac{3}{4}$ c	granulated sugar
6.	5 mL	1 tsp	vanilla extract
	2 eggs	2 eggs	eggs
7.	2 168 g packages	2 6 oz packages or 1 12 oz package	semisweet chocolate chips
8.	250 mL	1 c	chopped nuts

_____ 356°F _____ **9.** Preheat the oven to 180°C. Use $F = \frac{9}{5}C + 32$ to convert to Fahrenheit.

Sift flour, baking soda, and salt. In another bowl, cream butter, sugar, and vanilla. Beat in eggs. Add flour mixture 250 mL at a time; mix well. Stir in chocolate chips and nuts. Using a 5 mL measure, drop by rounded measures onto ungreased cookie sheet. Bake 8 to 10 minutes.

_____ 2 in. _____ **10.** Makes 100 (5 cm) cookies. How big are the cookies in inches?

Notice in the Swedish pancake recipe that some of the ingredients are in deciliters. While deciliter is not a commonly used measuring unit in the United States, it is often used in Europe. A deciliter is equivalent to 0.1 L, 10 cL, or 100 mL.

	Swedish Pancakes		
	Metric amount	**Customary amount**	**Ingredient**
11.	3 dL	$1\frac{1}{4}$ c	flour
12.	6 dL	$2\frac{1}{2}$ c	milk (divided)
	3 eggs	3 eggs	eggs
	0.3 mL	pinch	salt
13.	45 mL	9 tsp or 3 tbsp	butter

Combine the flour, 3 dL milk, the eggs, and salt in a bowl. Once those ingredients are well mixed, add the rest of the milk and mix well. Turn the heat on high and add about 5 mL butter to a skillet. When it is melted and sizzling, add about 1 dL of your mix to the pan. Lower the heat to medium high. Once the bottom of the pancake is cooked (2–3 minutes), flip it over and cook the other side. Serve with a dusting of powdered sugar and/or spread with jam. Makes about 9 pancakes.

Conversion Maze

Help Mr. Bee find his way through the hive to deposit the pollen he has gathered. Convert each measurement to determine the correct path. The first one is done for you as an example.

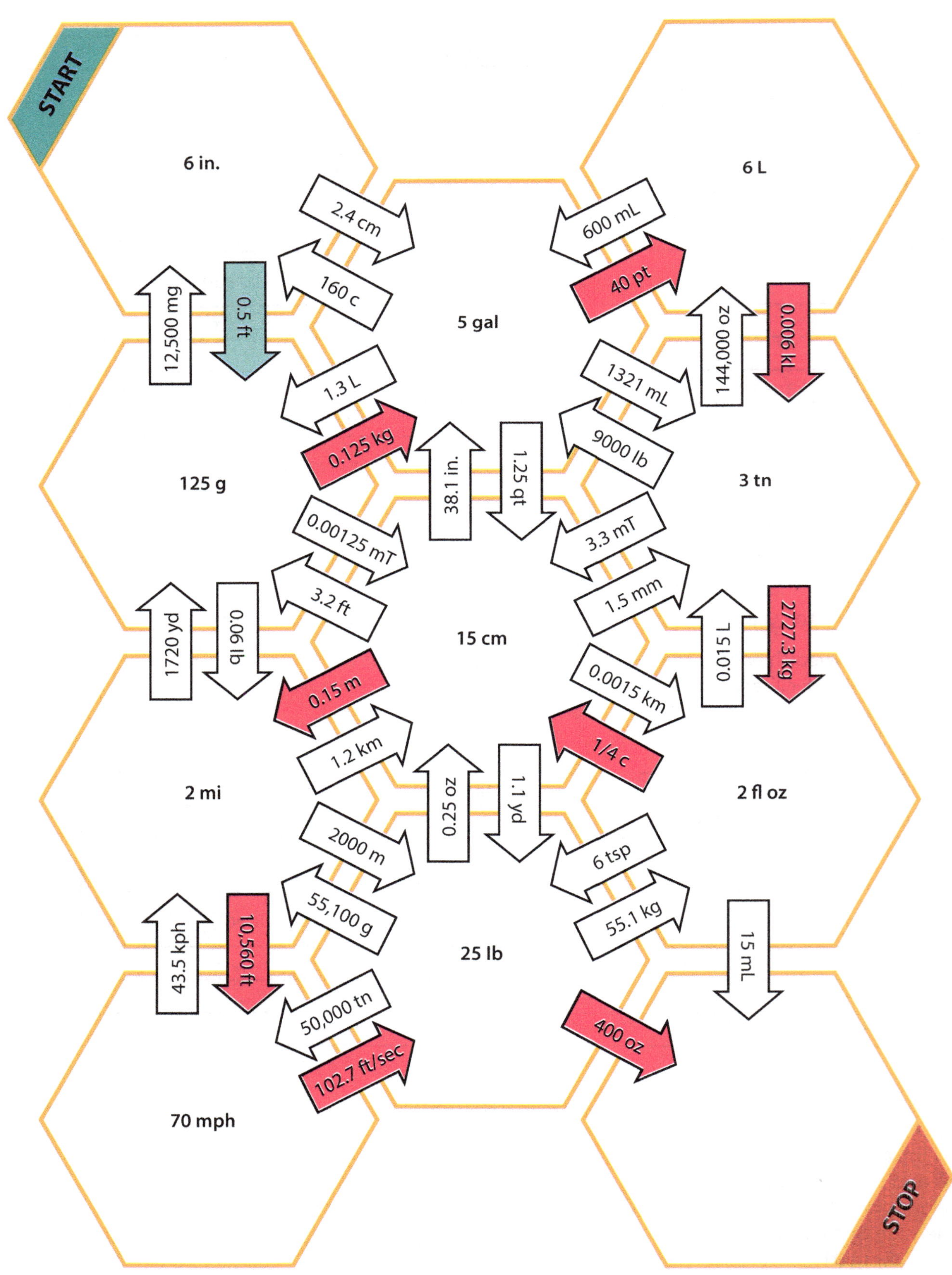

Fundamentals of Math

Pool Preparation

Name ____

1. Hallie's new pool is 3 m long, 6 m wide, and 1.5 m deep. Change the dimensions of the pool to centimeters.

   __________ cm long, __________ cm wide, and __________ cm deep 300, 600, 150

2. Find the number of cubic centimeters the pool holds. 27,000,000 cc

3. Convert cubic centimeters to liters. 27,000 L

4. If Hallie's hose can pump 8 gal per minute, how many hours will it take to fill her pool? (Round to the nearest whole hour.) 15 hr

5. How many cups of chlorine are needed for the pool if 1 c of chlorine powder treats 1000 gal of water? (Round to the nearest cup.) 7 c

Chapter 9 Cumulative Review

Name _______________________________________

_____A_____ **1.** Divide $6)\overline{295.2}$. **[1.4]**

    A. 49.2   **B.** 49.1

    C. 48.2   **D.** none of these

_____B_____ **2.** During the fall when the air is cooler and the water is warmer, evaporation from the Great Lakes can be as much as 0.6 in. per day. Write and solve a multiplication problem that models the change in lake levels in a week. **[2.5]**

    A. $0.6 \times (-7) = -4.2$ in.   **B.** $-0.6 \times 7 = -4.2$ in.

    C. $-0.6 \times (-7) = 4.2$ in.   **D.** $0.6 \times 7 = 4.2$ in.

_____B_____ **3.** Identify the property illustrated: $4 \times (12 + 5) = (4 \times 12) + (4 \times 5)$. **[3.3]**

    A. Associative   **B.** Distributive

    C. Commutative   **D.** none of these

_____A_____ **4.** Simplify the following expression: $(2t - 3) - (5t + 6)$. **[3.6]**

    A. $-3t - 9$ **B.** $3t - 3$

    C. $-3t + 3$ **D.** none of these

_____D_____ **5.** Which of the following are not multiples of the same number? **[4.3]**

    A. 14, 35, 70, 98, 105   **B.** 34, 68, 85, 102, 170

    C. 24, 72, 88, 104, 168   **D.** 38, 51, 57, 84

_____D_____ **6.** Which of the following is not rational? **[4.5]**

    A. $\sqrt{289}$ **B.** 4.56

    C. 37   **D.** all are rational

_____B_____ **7.** Divide $7\frac{7}{8} \div 3\frac{3}{4}$. **[5.5]**

    A. 2   **B.** $2\frac{1}{10}$

    C. $2\frac{3}{30}$ **D.** none of these

_____A_____ **8.** Find the correct solution to $2\frac{1}{3} + 3\frac{1}{2} \times \frac{3}{4}$. **[5.6]**

    A. $4\frac{23}{24}$ **B.** 7

    C. $4\frac{3}{8}$ **D.** none of these

_____B_____ **9.** Maggie's grandmother lives 490 mi away. If Maggie's dad drives an average of 65 mph on their trip to Grandma's house for Christmas, how long should the trip take? (Remember that *rate* × *time* = *distance*.) **[6.1]**

    A. about $6\frac{1}{2}$ hr   **B.** about $7\frac{1}{2}$ hr

    C. about 13 hr   **D.** about 31 hr

_____C_____ **10.** Solve for m: $3m + 21 = 9$. **[6.4]**

    A. $m = 10$ **B.** $m = 7$

    C. $m = -4$ **D.** none of these

_____C_____ **11.** Translate the following sentence into an algebraic inequality: "The difference between a number and seven is at least 52." **[6.6]**

    A. $n - 7 \leq 52$ **B.** $7 - n \leq 52$

    C. $n - 7 \geq 52$ **D.** $7 - n \geq 52$

_____B_____ **12.** Becca's recipe makes 5 dozen chocolate chip cookies. The recipe calls for 12 oz of chocolate chips. She has read that there are about 54 chocolate chips in an ounce. What, then, should the unit rate of chocolate chips to cookies be? **[7.1]**

    A. 5 chips to 1 cookie   **B.** 10.8 chips to 1 cookie

    C. 129.6 chips to 1 cookie   **D.** none of these

_____C_____ **13.** In Rachelle's youth orchestra there are 8 violinists and 3 violists. If a professional orchestra has the same ratio of violins to violas and there are 12 violists, how many violinists are there? **[7.3]**

    A. 12 violinists   **B.** 24 violinists

    C. 32 violinists   **D.** none of these

_____C_____ **14.** Alex knows that the constant of proportionality in the ratio of boys to pizza slices is 4.5. How many pizza slices should he be sure to have if he has invited 5 friends over to play video games with him? (Alex will be eating too!) **[7.5]**

    A. 1.3 slices   **B.** 22.5 slices

    C. 27 slices   **D.** not enough information

_____B_____ **15.** Change this fraction to a percent: $\frac{3}{5}$. **[8.1]**

    A. 0.6%   **B.** 60%

    C. 167%   **D.** none of these

_____C_____ **16.** Find the sales tax of 6.5% on an item costing $69.95. **[8.2]**

    A. $4.54   **B.** $4.75

    C. $4.55   **D.** none of these

_____B_____ **17.** Enlarge a 5×7 in. picture to 140% of its original size. **[8.5]**

    A. 7×9 in.   **B.** 7×9.8 in.

    C. 7.8×9.8 in.   **D.** none of these

_____C_____ **18.** If a $47 sweater is on sale for 25% off, what will the sale price be? **[8.7]**

    A. $36   **B.** $36.75

    C. $35.25   **D.** none of these

Measuring Angles
(Extra Practice; use after Section 10.1)

Name _______________________

Measure each angle and classify it as right, acute, obtuse, or straight.

__20°; acute__ **1.**

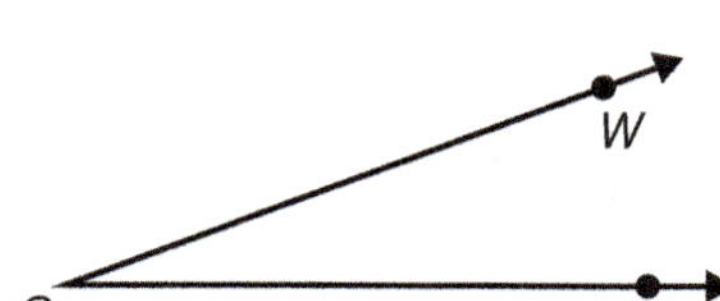

__110°; obtuse__ **2.**

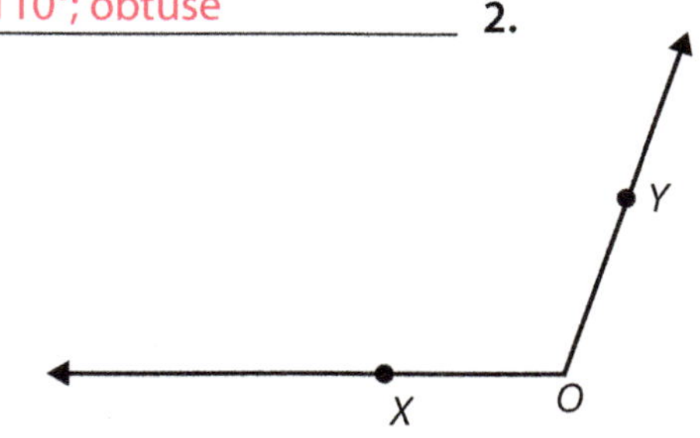

__20°; acute__ **3.**

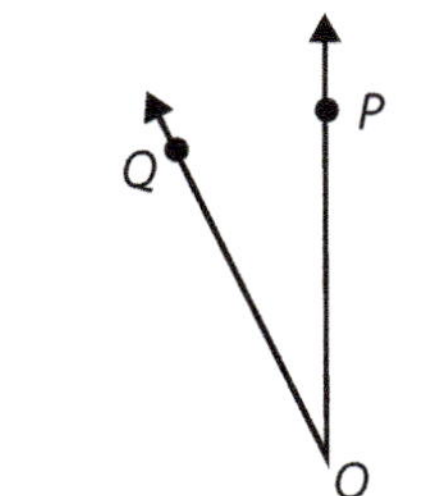

__55°; acute__ **4.**

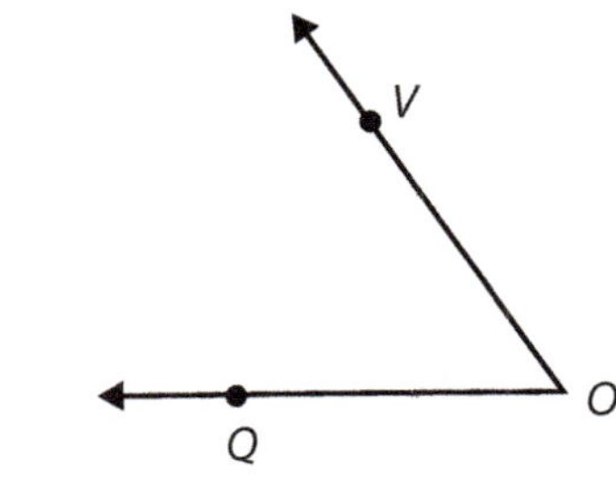

__90°; right__ **5.**

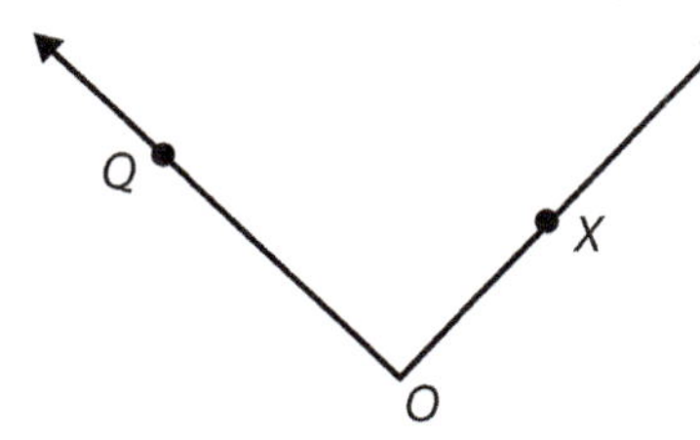

__44°; acute__ **6.**

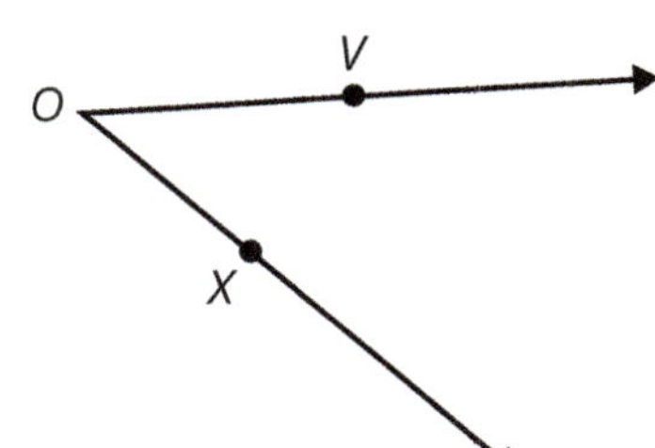

__65°; acute__ **7.**

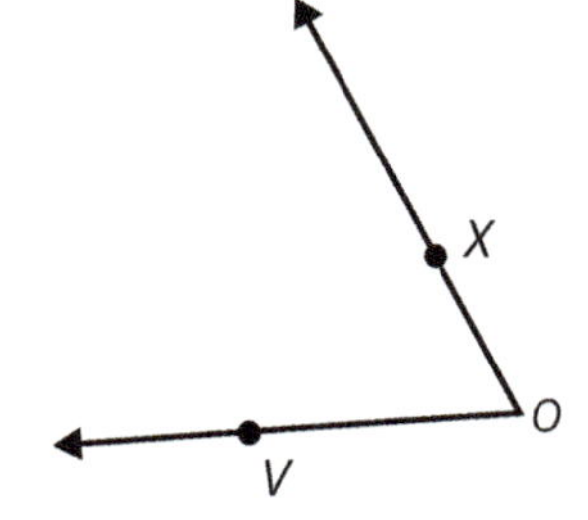

__155°; obtuse__ **8.**

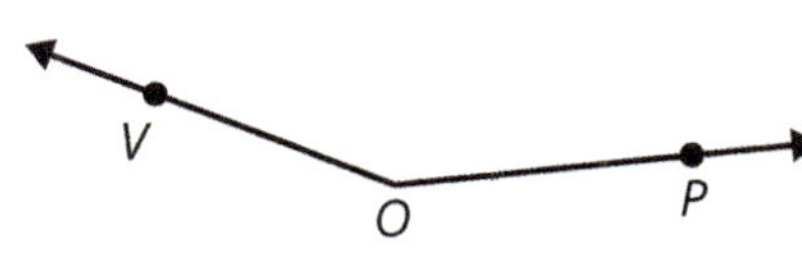

__135°; obtuse__ **9.**

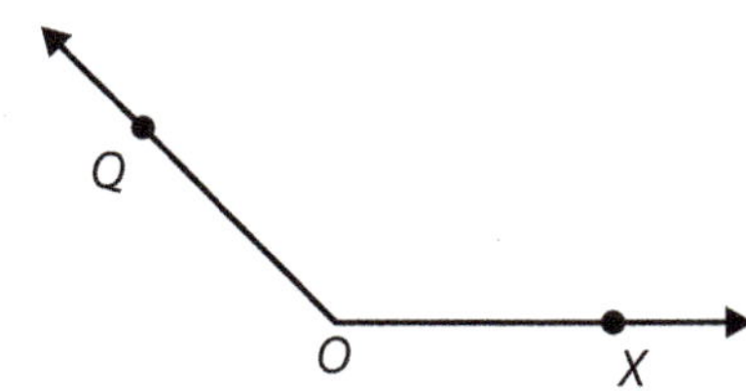

__120°; obtuse__ **10.**

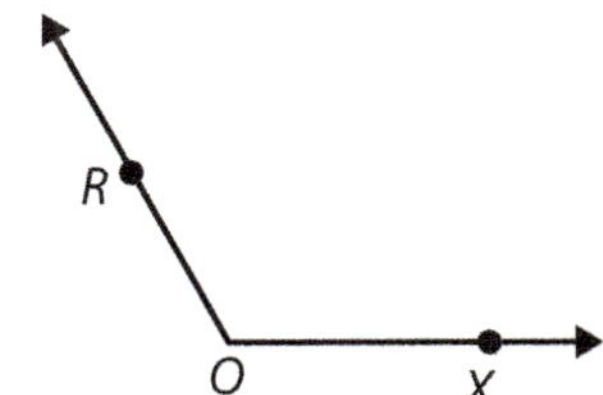

11. $\angle CAB = 30°$

12. $\angle DAB = 90°$

13. $\angle EAB = 150°$

14. $\angle FAB = 180°$

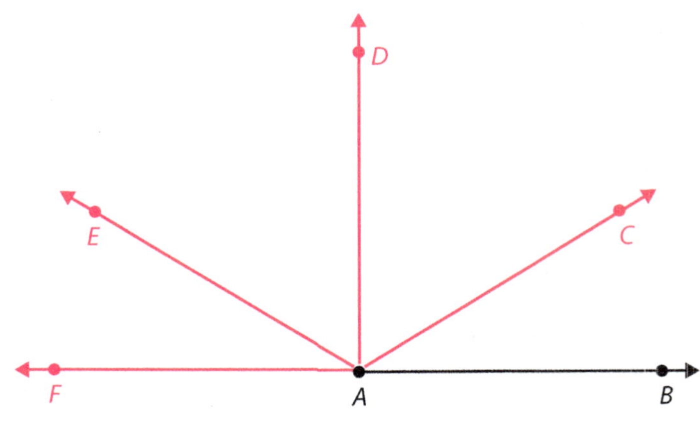

15. What is the measure of $\angle EAC$? 120°

Lines and Angles

Name _______________________________

Example: Given $m\angle 1 = 43°$, find $m\angle 2$.

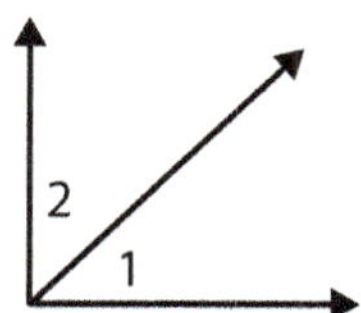

Answer:

If $m\angle 1 = 43°$, then $m\angle 2 = 47°$, since $90 - 43 = 47$.
Reason: $\angle 2$ is a complement of $\angle 1$.

Give a reason for each answer in exercises 1–8.

1. Given $m\angle 1 = 152°$, find $m\angle 2$. 28°; reason: $\angle 2$ is a supplement of $\angle 1$.

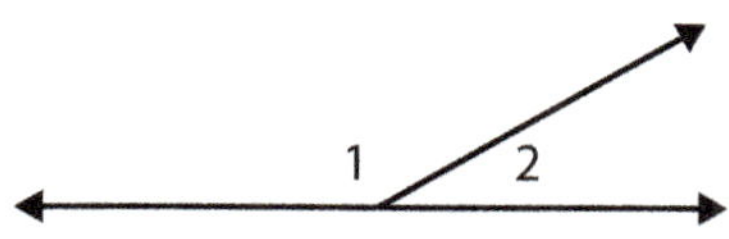

Use the given figure for exercises 2–4. If $m\angle 1 = 32°$, then

2. find $m\angle 2$. 148°; reason: $\angle 2$ is a supplement of $\angle 1$.

3. find $m\angle 3$. 32°; reason: $\angle 3$ and $\angle 1$ are vertical angles.

4. find $m\angle 4$. 148°; reason: $\angle 4$ and $\angle 2$ are vertical angles.

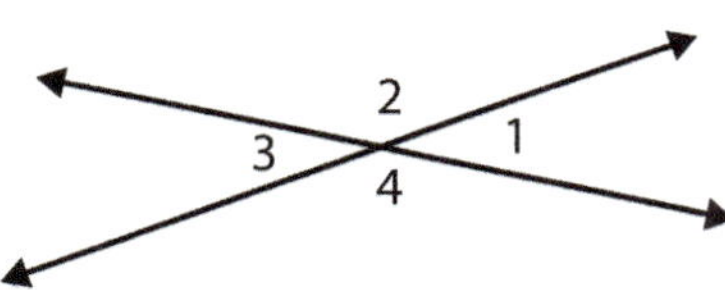

Use the given figure for exercises 5–8. If $m\angle CFD = 38°$, then

5. find $m\angle BFC$. 52°; reason: $\angle BFC$ is a complement of $\angle CFD$.

6. find $m\angle AFE$. 90°; reason: $\angle AFE$ is a supplement to $\angle AFB$.

7. find $m\angle EFD$. 90°; reason: $\angle EFD$ is a supplement to $\angle AFE$.

8. find $m\angle AFC$. 142°; reason: $\angle AFC$ is a supplement to $\angle CFD$.

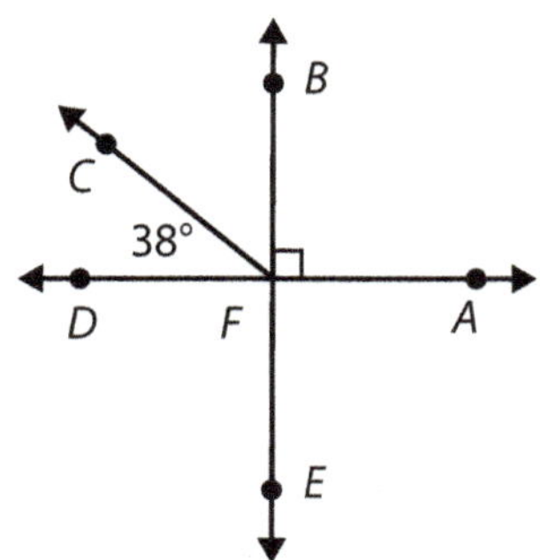

Triangles

Name ______________________________

Classify each triangle two ways.

right; scalene   **1.**

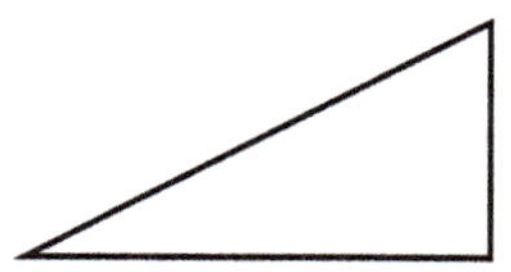

acute; isosceles   **2.**

obtuse; scalene   **3.**

obtuse; scalene   **4.**

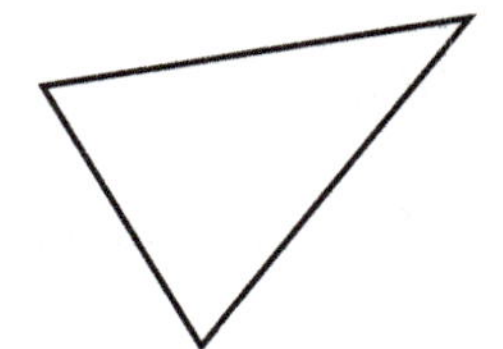

acute; scalene   **5.**

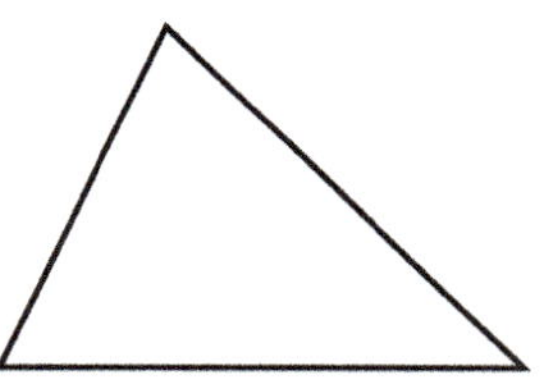

obtuse; isosceles   **6.**

acute; scalene   **7.**

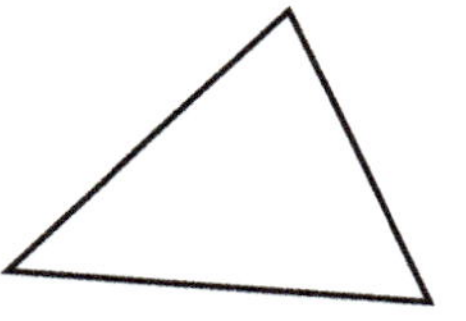

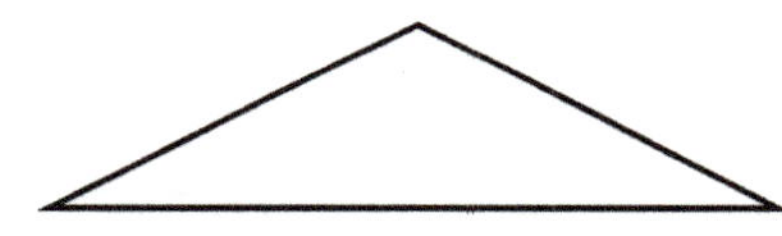

right; scalene   **8.** a triangle containing a 10° and an 80° angle

acute; equilateral   **9.** a triangle with 2 angles measuring 60° each

obtuse; scalene   **10.** a triangle with a 40° and a 30° angle

right; isosceles   **11.** a triangle with two 45° angles

Write the measure of the missing angles.

<u> 90°; 50° </u> **12.** $\angle A$ and $\angle B$ if $\angle C$ is a 40° angle in right $\triangle ABC$

<u> 20° each </u> **13.** $\angle F$ and $\angle G$ if $\angle H$ is a 140° angle in obtuse isosceles $\triangle FGH$

<u> 60° each </u> **14.** $\angle X$ and $\angle Y$ if $\angle Z$ is a 60° angle in acute equilateral $\triangle XYZ$

<u> 45° each </u> **15.** $\angle R$ and $\angle T$ if $\angle S$ is a right angle in isosceles $\triangle RTS$

Finding Composite Area

Name ___________________________

Jasmine is pricing a few things based on the floor plans for her new condo.

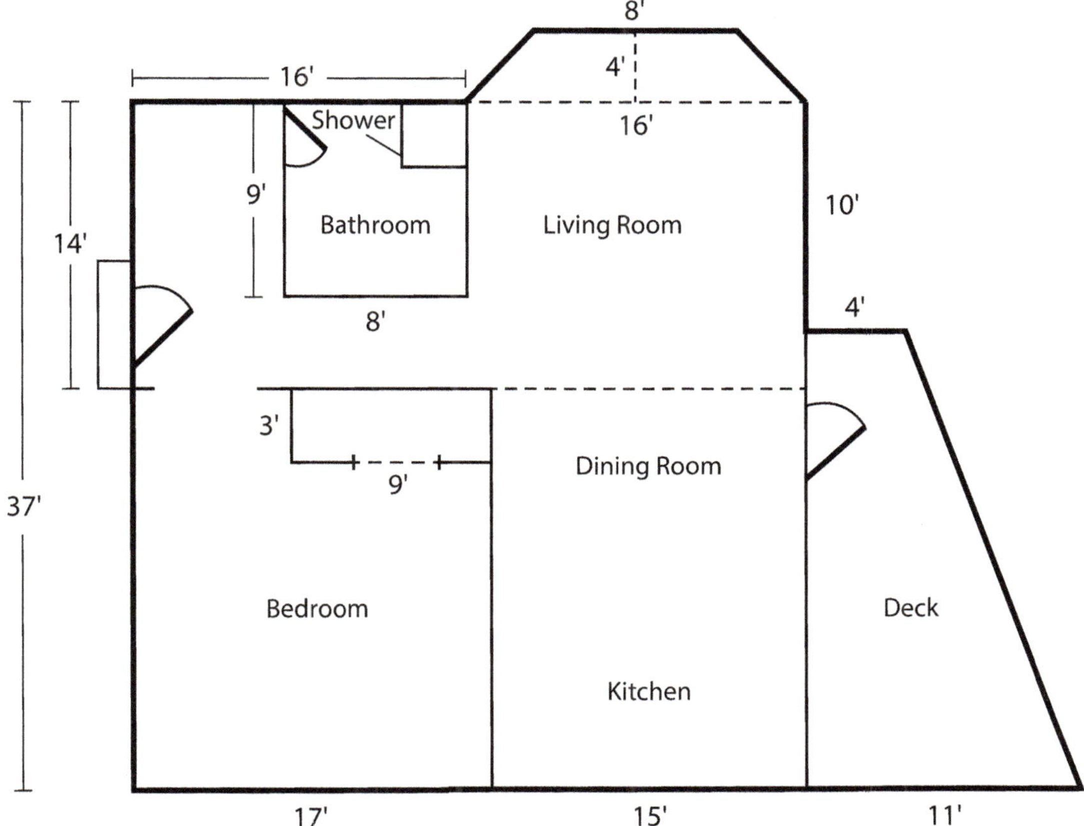

__345 ft²__   **1.** How many square feet is the kitchen/dining room?

__$1518__   **2.** How much will she pay for the kitchen floor if the tile costs $4.40 per square foot installed?

__63 ft²__   **3.** How many square feet is the bathroom, not counting the 3 × 3 shower?

__$277.20__   **4.** How much will the bathroom floor cost if she uses the same flooring as the kitchen?

__202.5 ft²__   **5.** How many square feet is the deck?

__$60.75__   **6.** What is the total price to stain the deck if it costs $0.30 per square foot?

__815 ft²__   **7.** What is the total square footage of the entryway, living room, and bedroom?

__$5256.75__   **8.** If hardwood flooring costs $6.45 per square foot installed, what will the total cost be if she puts hardwood flooring in the entryway, living room, and bedroom?

__$7112.70__   **9.** Find the total cost for the tile, stain, and hardwood.

Construction of Triangles

Name ________________________________

Tools needed: straightedge, protractor, compass if available, and your own paper (instructions given for working with and without a compass)

Construct a triangle with given angles.

The 2 angles we know are 30° and 60°. (What is the 3rd angle?)   90°

1. Using your straightedge, draw a line segment at the bottom of the space where you want to draw the triangle.

2. Using your protractor, make a 30° angle at the left edge of your line segment.

3. Using your protractor, make a 60° angle at the right edge of your segment.

4. The point where the 2 lines cross is the 3rd vertex of your triangle. Measure this angle to verify it is what you were expecting.

5. Could these instructions produce 1 unique triangle, more than 1 triangle, or no triangles?   1 unique triangle

Construct a triangle given 1 side and 2 angles.

The side we know is shown below, and the angles are 100° and 30°. (What is the 3rd angle?)   50°

____

1. Using your straightedge, draw a line segment at the bottom of your triangle space. Mark the length of the line segment using either a ruler or a compass.

2. Using your protractor, make a 100° angle at the left edge of your line segment.

3. Using your protractor, make a 30° angle at the right edge of your segment.

4. The point where the 2 lines cross is the 3rd vertex of your triangle. Measure this angle to verify it is what you were expecting.

5. Could these instructions produce 1 unique triangle, more than 1 triangle, or no triangles?   1 unique triangle

Construct a triangle given 2 sides and 1 angle.

The 2 sides we know are shown below, and the angle is 45°.

____

____

1. Using your straightedge, draw a line segment at the bottom of your triangle space. Mark the length of the longer line segment using either a ruler or a compass.

2. Using your protractor, make a 45° angle at the left edge of your line segment.

3. Mark the length of the shorter line segment on the upper arm of the angle.

4. Using your straightedge, connect the ends of the 2 line segments.

5. Could these instructions produce 1 unique triangle, more than 1 triangle, or no
triangles?   1 unique triangle

Construct a triangle given 3 sides (requires a compass).
The 3 sides we know are shown below.

____
____
____

1. Using your straightedge, draw a line segment at the bottom of your triangle space.
   Mark the length of the longer line segment using either a ruler or a compass.

2. Put the point of your compass on the end of the shortest line segment, and then open
   it up so that the tip of the pencil is on the other end of the segment.

3. Without changing the width of the compass arms, move the point to the left end of
   the long segment in your drawing. Draw an arc with the pencil in the space above the
   longer segment.

4. Put the point of your compass on the end of the middle line segment, and then open
   it up so that the tip of the pencil is on the other end of the segment.

5. Without changing the width of the compass arms, move the point to the right end of
   the long segment in your drawing. Draw an arc with the pencil in the space above the
   longer segment.

6. The point where the 2 arcs cross is the 3rd vertex of your triangle. Using your
   straightedge, connect the vertex to the endpoints of the longer segment.

7. Could these instructions produce 1 unique triangle, more than 1 triangle, or no
   triangles?   1 unique triangle

Repeat the above instructions with 2 more sets of lines.
The 3 sides we know are shown below.

____
____
____

8. Could these instructions produce 1 unique triangle, more than 1 triangle, or no
   triangles?   no triangles

The 3 sides we know are shown below.

____
____
____

9. Could these instructions produce 1 unique triangle, more than 1 triangle, or no
   triangles?   1 unique triangle

Geometry Terms

(Mixed Practice; use after Section 10.6)

Name _______________________

Find and circle the words defined below. The words may be found written vertically, horizontally, or diagonally.

p	m	v	e	r	t	e	x	j	a	r	f	l	s	l
t	p	p	g	f	r	l	d	o	u	h	q	i	q	r
c	r	p	a	e	y	r	n	p	b	q	v	n	u	h
f	o	a	e	r	o	l	e	n	o	t	g	e	a	o
e	t	n	p	r	a	m	f	c	s	l	u	e	r	m
q	r	s	g	e	p	l	e	r	t	y	y	s	e	b
u	a	a	c	r	z	e	l	t	i	a	v	g	e	u
i	c	c	r	a	u	o	n	e	r	i	n	g	o	s
l	t	u	a	s	l	e	i	d	l	y	n	g	b	n
a	o	t	y	d	i	e	n	d	i	o	d	o	l	f
t	r	e	u	i	z	y	n	t	a	c	g	i	n	e
e	i	s	o	s	c	e	l	e	s	x	u	r	e	l
r	k	v	u	h	h	p	h	c	g	q	t	l	a	y
a	t	s	t	r	a	i	g	h	t	q	j	e	a	m
l	r	v	i	m	d	c	u	c	r	i	g	h	t	r

1. the common endpoint of the sides of an angle
2. a closed figure made up of line segments
3. a triangle with all equal sides
4. an angle measuring 180°
5. a quadrilateral with opposite sides parallel
6. intersecting lines that form 90° angles
7. a quadrilateral with exactly 1 pair of parallel sides
8. a parallelogram with equal sides
9. a triangle with at least 2 equal sides
10. the study of shapes
11. an angle measuring more than 90° but less than 180°
12. an instrument used to measure angles
13. same shape and size
14. a rhombus with 4 right angles
15. a parallelogram with right angles
16. a part of a line extending infinitely from a point
17. the type of triangle needed to apply the Pythagorean Theorem
18. a triangle with no equal sides
19. a set of points along a straight path with no endpoints
20. a triangle with all 3 angles less than 90°

Chapter 10 Cumulative Review

B   **1.** Place the decimal point to make the correct product: $0.007 \times 0.38 = 266$. **[1.3]**

     A. 0.266   **B.** 0.00266

     C. 0.0266   **D.** none of these

D   **2.** Simplify $-2 + (-5.7)$ **[2.3]**

     A. -5.9 **B.** -3.7

     C. 7.7   **D.** none of these

B   **3.** Which of the following is equivalent to the expression $7x - 42$? **[3.4]**

     A. $7(x - 42)$ **B.** $7(x - 6)$

     C. $6(x - 7)$ **D.** none of these

D   **4.** Find the greatest common factor of 68 and 99. **[4.2]**

     A. 2   **B.** 17

     C. 3   **D.** none of these

D   **5.** List the following fractions in increasing order: $\frac{13}{16}, \frac{3}{4}, \frac{17}{20}, \frac{7}{9}$. **[4.6]**

     A. $\frac{13}{16}, \frac{3}{4}, \frac{7}{9}, \frac{17}{20}$ **B.** $\frac{3}{4}, \frac{13}{16}, \frac{17}{20}, \frac{7}{9}$

     C. $\frac{3}{4}, \frac{13}{16}, \frac{7}{9}, \frac{17}{20}$ **D.** none of these

C   **6.** Subtract $\frac{7}{9} - \left(-\frac{5}{6} \right)$. **[5.2]**

     A. $\frac{2}{3}$ **B.** $-\frac{1}{18}$

     C. $1\frac{11}{18}$ **D.** none of these

A   **7.** Multiply $3\frac{3}{5} \cdot 4\frac{1}{6}$. **[5.4]**

     A. 15   **B.** $12\frac{1}{10}$

     C. $\frac{1}{15}$ **D.** none of these

B   **8.** Solve for x: $1.5x = 7.5$. **[6.3]**

     A. $x = 6$ **B.** $x = 5$

     C. $x = 3$ **D.** none of these

C   **9.** Solve for x: $3 - 4x \leq -25$. **[6.8]**

     A. $x \leq 7$ **B.** $x \leq 5.5$

     C. $x \geq 7$ **D.** $x \geq 5.5$

A   **10.** Solve the following proportion: $\frac{x}{12} = \frac{17}{54}$. Round to the nearest tenth. **[7.3]**

     A. 3.8   **B.** 4.0

     C. 85.1   **D.** none of these

A   **11.** Emma and Bethany are having a contest to see how fast they can do math facts. Emma did a page of 28 problems in 77 seconds, and Bethany did a page of 35 problems in 100 seconds. Which girl was working faster? **[7.1]**

     A. Emma   **B.** Bethany

     C. same rate   **D.** impossible to say

_____**12.** On a scale drawing, a segment $4\frac{3}{4}$ ft represents 190 mi. What is the scale? **[7.4]**

A. 1 in. : 60 mi

B. 1 in. : 55 mi

C. 1 in. : 125 mi

D. none of these

_____**13.** Find the amount earned with a 7% commission on $3857.20 in sales. **[8.1]**

A. $270

B. $270.40

C. $270.04

D. none of these

_____**14.** What percent of 68 is 25.5? Round to the nearest whole percent. **[8.4]**

A. 267%

B. 27%

C. 38%

D. none of these

_____**15.** The attendance increased from 8000 the first year to 10,800 the second year. Find the percent increase to the nearest percent. **[8.6]**

A. 26%

B. 74%

C. 35%

D. none of these

_____**16.** Change 34 yd to inches. **[9.1]**

A. 408 in.

B. 1224 in.

C. 1020 in.

D. none of these

_____**17.** Change 10 qt to gallons. **[9.2]**

A. 2.5 gal

B. 3.3 gal

C. 5 gal

D. none of these

_____**18.** Change 300 m to kilometers. **[9.3]**

A. 3 km

B. 0.3 km

C. 30 km

D. none of these

_____**19.** Knowing that there are about 2.2 lb in a kilogram, change 25 lb to kilograms. **[9.4]**

A. 0.88 kg

B. 11.36 kg

C. 55 kg

D. none of these

_____**20.** Natalie ran a 2 mi race in 18 min. Find her rate in feet per second. **[9.5]**

A. 9.78 ft/sec

B. 13.2 ft/sec

C. 586.67 ft/sec

D. none of these

Exploring Area and Volume

Areas of Circles
(Mixed Practice; use after Section 11.1)

Name ________________________

Consider the circles below with center O. These are called *concentric circles* because they share a common center point. $AH = 16$ cm, $BG = 12$ cm, $CF = 10$ cm, and $DE = 4$ cm.

Find the length of the radius of the circle containing the given point.

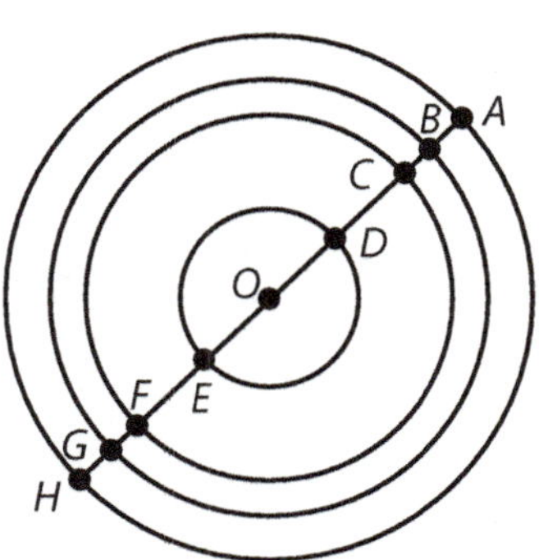

_8 cm_ **1.** A

_6 cm_ **2.** B

_5 cm_ **3.** C

_2 cm_ **4.** D

Find the area of the circle containing the given point. Use 3.14 for π.

12.56 cm² **5.** D

78.5 cm² **6.** C

113.04 cm² **7.** B

200.96 cm² **8.** A

Find the area of the shaded figure. Use 3.14 for π. Round to the nearest hundredth.

29.03 mm² **9.**

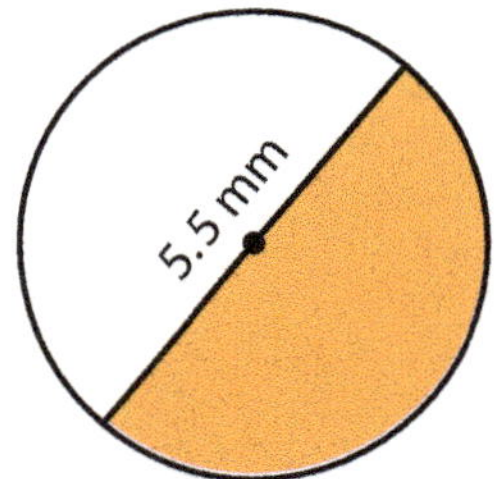

_24.53 m²_ **10.**

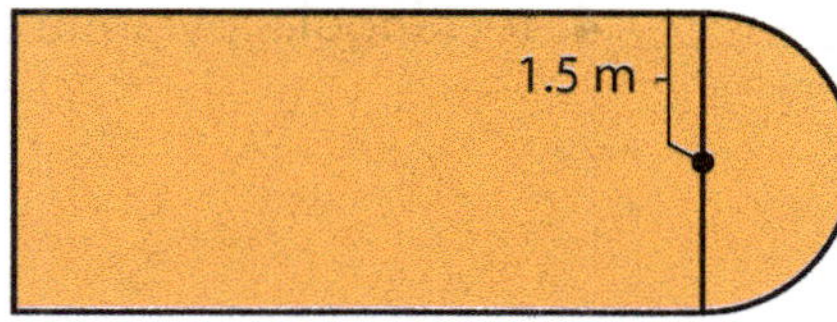

Three-Dimensional Figures

(Extra Practice; use after Section 11.4)

Draw each of the following 3-dimensional figures. Then, give the number of faces (*F*), edges (*E*), and vertices (*V*).

$F = 6; E = 12; V = 8$ **1.** cube

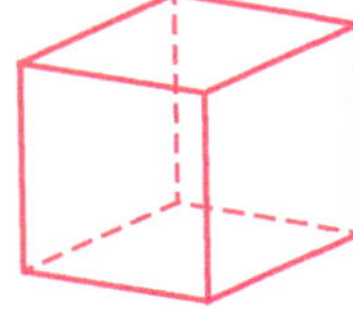

$F = 4; E = 6; V = 4$ **2.** triangular pyramid

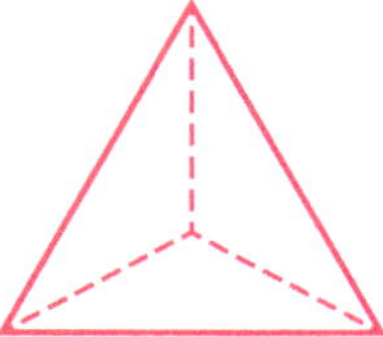

$F = 5; E = 9; V = 6$ **3.** triangular prism

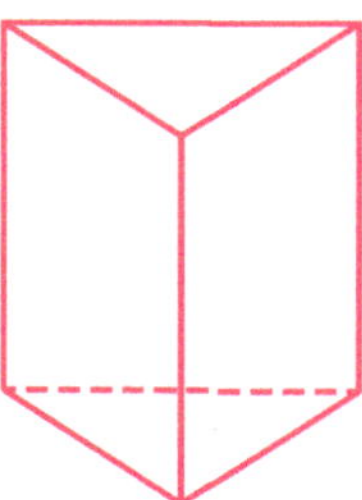

$F = 5; E = 8; V = 5$ **4.** rectangular pyramid

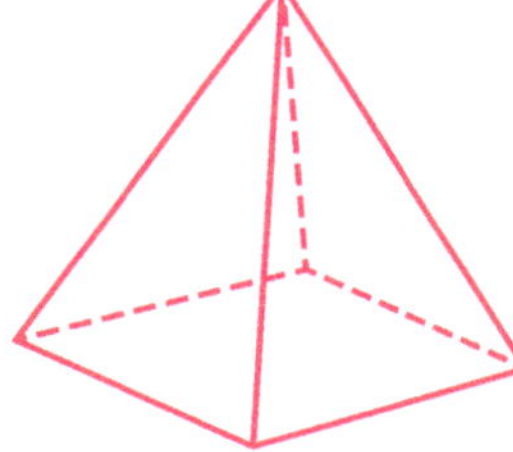

$F = 7; E = 15; V = 10$ **5.** pentagonal prism

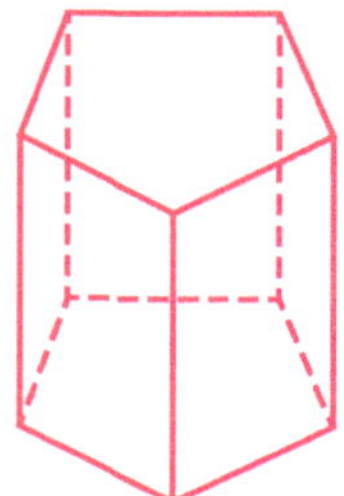

6. Can you see a pattern to the number of faces, edges, and vertices in a prism that would help you to predict (without drawing) the number of each there would be in an 18-gon prism? Faces = sides + 2; Edges = sides × 3; Vertices = sides × 2, so for an 18-gon $F = 20; E = 54; V = 36$

7. What about an 18-gon pyramid? Faces = sides + 1; Edges = sides × 2; Vertices = sides + 1, so for an 18-gon $F = 19; E = 36; V = 19$

Surface Area

(Extra Practice; use after Section 11.4)

Name _______________________________

Find the surface area.

___290 cm²___ **1.**

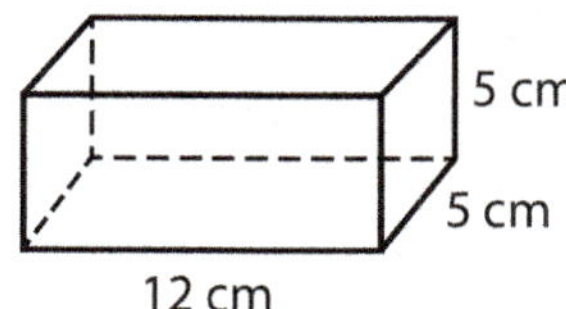

___429 mm²___ **2.**

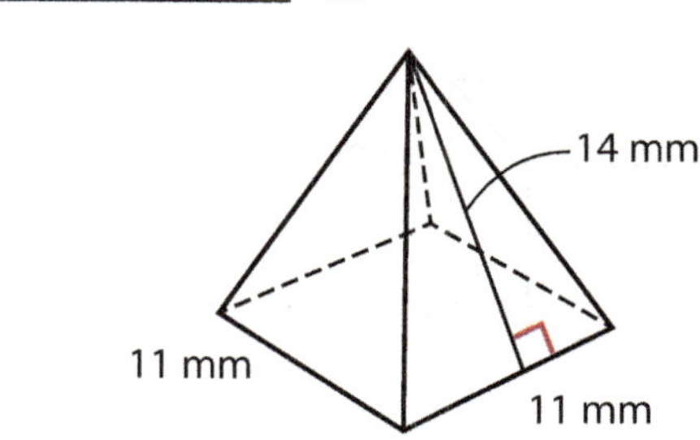

___123.2 in.²___ **3.**

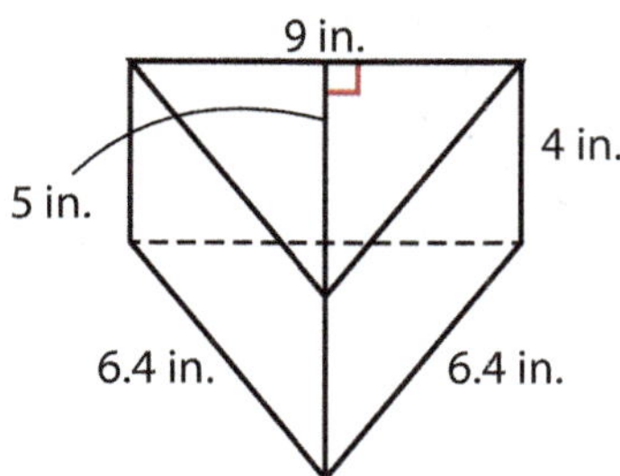

___35.06 cm²___ **4.**

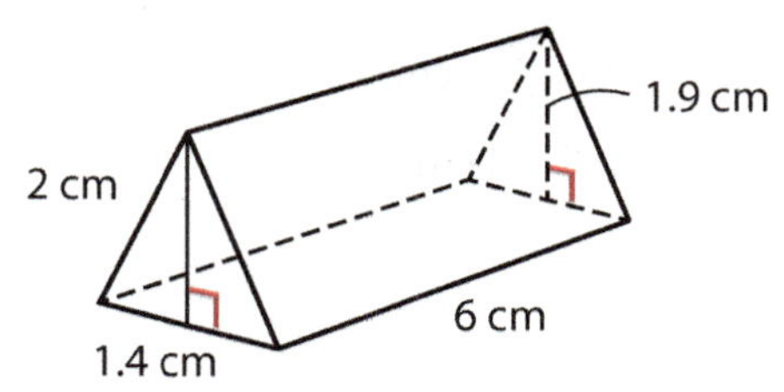

___37.6 cm²___ **5.**

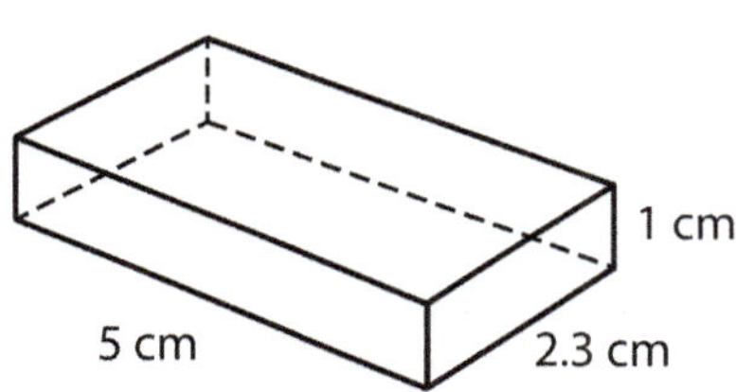

___125 in.²___ **6.**

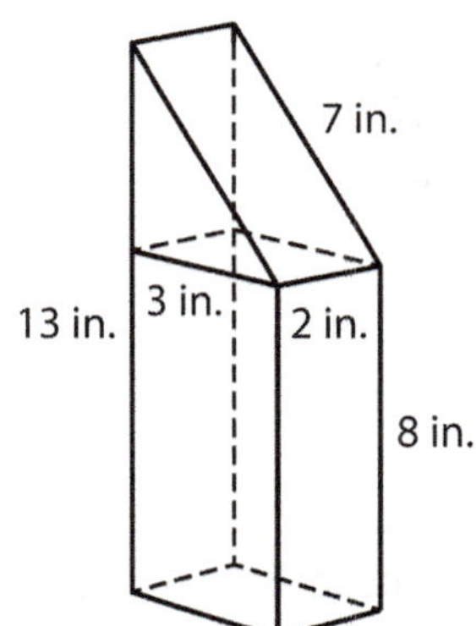

___5400 cm²___ **7.** A cube measures 30 cm on each edge. What is its surface area?

___1832 cm²___ **8.** A rectangular prism has a length of 22 cm, a width of 13 cm, and a height of 18 cm. What is its surface area?

Volume

Find the volume of each prism.

<u>300 cm³</u> **1.**

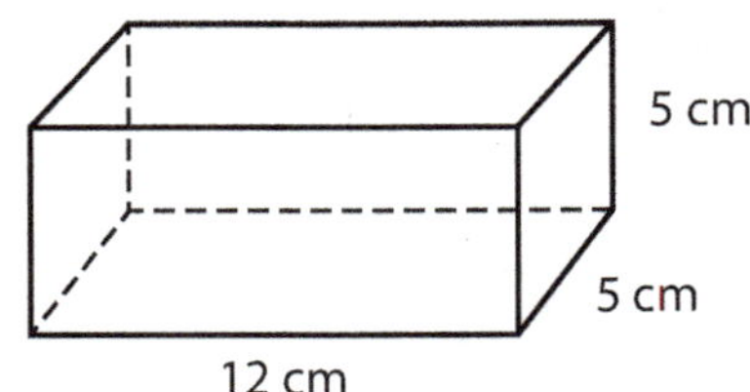

<u>80 in.³</u> **2.**

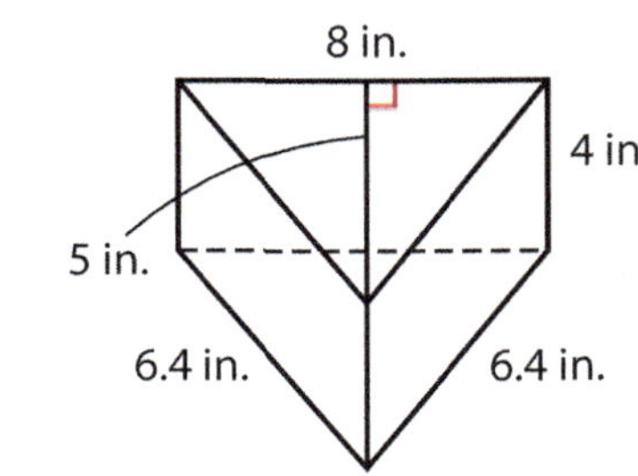

<u>11.5 cm³</u> **3.**

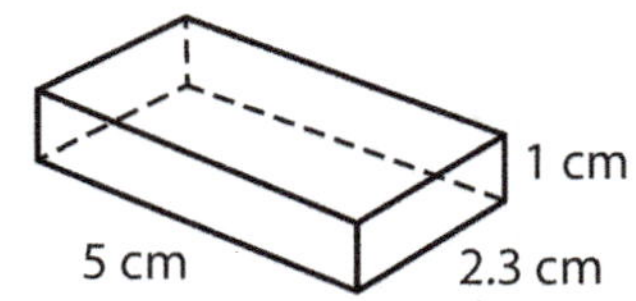

<u>162.225 mm³</u> **4.**

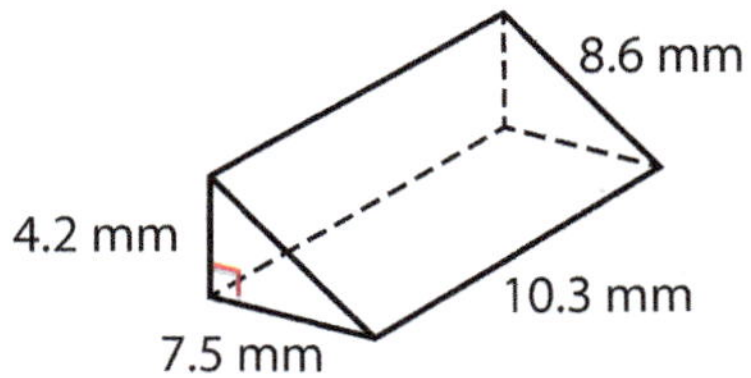

<u>7.98 cm³</u> **5.**

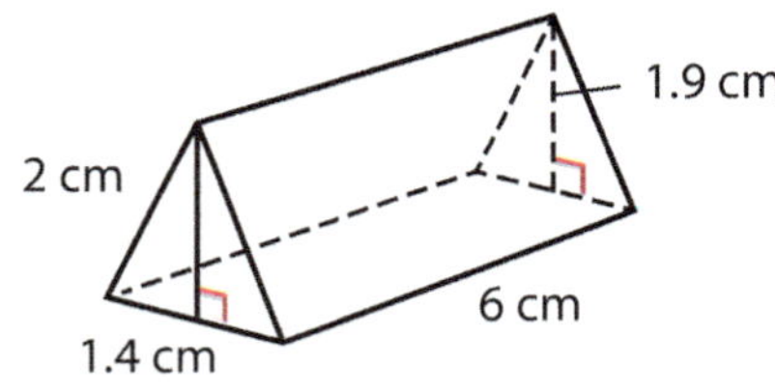

<u>63 in.³</u> **6.**

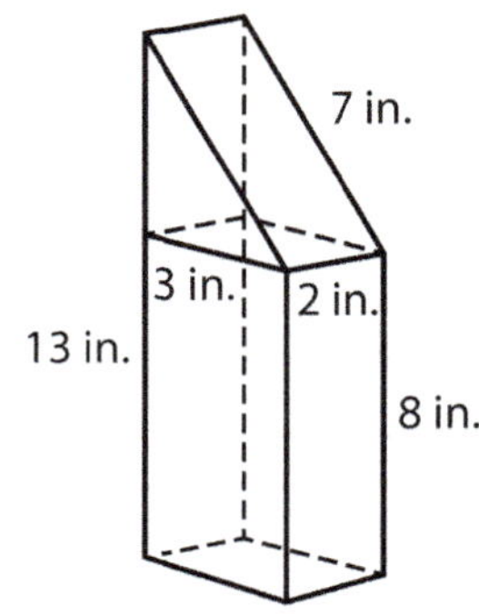

<u>1350 cm³</u> **7.** area of base: 54 cm²; height: 250 mm

<u>2700 cm³</u> **8.** area of base: 150 cm²; height: 0.18 m

Solve.

<u>20,520 cm³</u> **9.** The drawer in Ashley's dresser is 45 cm wide, 38 cm deep, and 12 cm high. What is the volume of the drawer?

<u>9072 cm³</u> **10.** The drawer in Andrew's desk is 28 cm wide, 54 cm deep, and 6 cm high. What is the volume of the drawer?

Fundamentals of Math

Cross Sections of 3D Shapes

(Enrichment; use after Section 11.5)

Name ______________________________

You will need several things for this activity.

- A cylinder—use a toilet paper or paper towel tube. Depending on how much sand you have, you might want to cut this off to be about 2 in. high.

- A rectangular prism—any small empty box will do.

- A pyramid—cut 4 copies of an isosceles triangle out of thin cardboard or heavy cardstock; then tape them together with duct tape or some other heavy tape.

- Kinetic sand—this can be purchased, or you can make your own with this recipe.

- A piece of thin cardboard or heavy cardstock to use as a slicer—this will be referred to as your "plane."

> **Kinetic Sand Recipe**
>
> 1 c play sand   1 Tbsp dish soap
>
> 1 Tbsp corn starch   1 c water
>
> 1. Combine the sand and corn starch in a large container.
>
> 2. In a measuring cup, combine water and dish soap. Stir till bubbly.
>
> 3. Slowly pour the water mixture into the sand mixture, mixing as you go. You may not need the whole amount or you may need a little extra water. Continue until you reach the desired consistency.
>
> 4. Store in an airtight container.

Activity Instructions

1. Make a sand cylinder using your cylinder mold. Cut your model using your plane in the following ways and describe the shape you see.

    a. Slice the model holding your plane parallel to the base. What shape is the top of the model now? a circle

    b. Slice the model holding your plane perpendicular to the base, cutting along the diameter. What shape is the front of the model now? a rectangle

    c. Slice the remaining part of the model in half. What shape is the front of the model now? What is the relationship between this shape and the previous shape? a rectangle (same height, smaller width)

    d. Recreate your cylinder and slice it again but this time at a diagonal. What shape is the top of the model now? an oval

2. Make a rectangular prism using your prism mold. Cut your model using your plane in the following ways and describe the shape you see.

    a. Slice the model holding your plane parallel to the base. What shape is the top of the model now? a rectangle

    b. Slice the model, holding your plane perpendicular to the base. What shape is the front of the model now? a rectangle

    c. Slice the remaining part of the model in half. What shape is the front of the model now? What is the relationship between this shape and the previous shape? a rectangle (congruent)

3. Make a square pyramid using your pyramid mold. Cut your model using your plane in the following ways and describe the shape you see.

   a. Slice the model, holding your plane parallel to the base. What shape is the top of the model now? a square

   b. Slice the model, holding your slicer perpendicular to the base, cutting straight down through the apex. What shape is the front of the model now? a triangle

   c. Slice the remaining part of the model in half. What shape is the front of the model now? a trapezoid

4. Use your plane to shape your sand as nearly as possible into a perfect cube. Can you figure out how to use your plane to slice this figure into each of the following shapes?

   a. square

   b. rectangle

   c. triangle

   d. hexagon

Derived Areas and Volumes

Name ________________________________

Many real-life figures are made up of several of the basic shapes, such as rectangles, squares, and circles. Likewise, many solids are composed of a combination of several basic objects such as cylinders and prisms. To find areas or volumes, we sometimes need to add or subtract the areas or the volumes of several basic figures.

Example: A circular flower bed has a diameter of 10 ft, with a 3-foot-wide circular walk around the outside made of brick pavers. Knowing there are 4.5 pavers per square foot, find the number of pavers used by finding the area of the walk.

Answer:

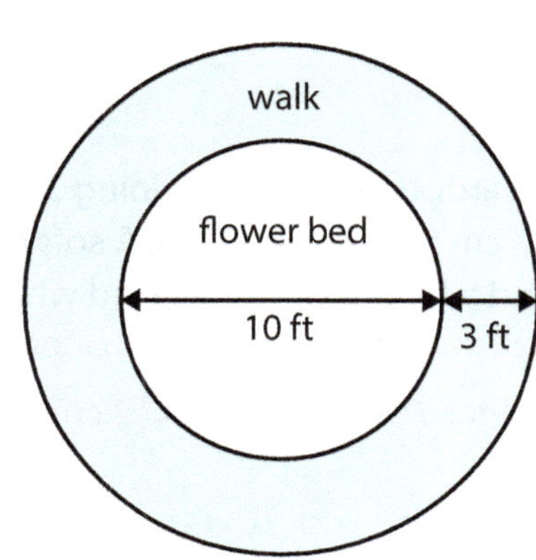

Radius of outer circle: $5 + 3 = 8$ ft

Area of outer circle: $\pi(8^2) = 64\pi = 64(3.14) = 200.96$ ft^2

Radius of inner circle: 5 ft

Area of inner circle: $\pi(5^2) = 25\pi = 25(3.14) = 78.5$ ft^2

Area of walk: $200.96 - 78.5 = 122.46$ ft^2

Number of pavers: $4.5(122.46) = 551.07 \approx 552$ pavers

Find the area of the given derived figures.

1. semicircular casing above a round window
   $0.5[\pi 22^2) - \pi(18^2)] \approx 251.2$ in.^{2}

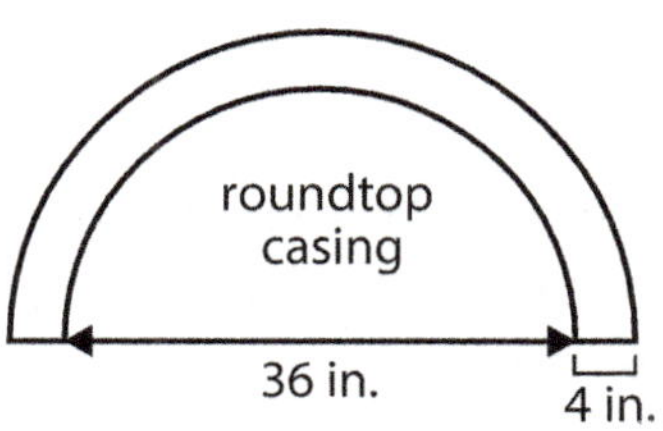

2. free-throw zone on a basketball court
   $12(18) + 0.5[\pi(6^2)] \approx 272.52$ ft^2

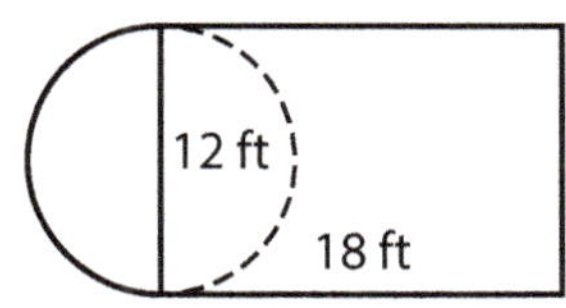

3. base of an advertising kiosk
   $4[0.5\pi(1)^2] + 2^2 \approx 10.28$ m^2

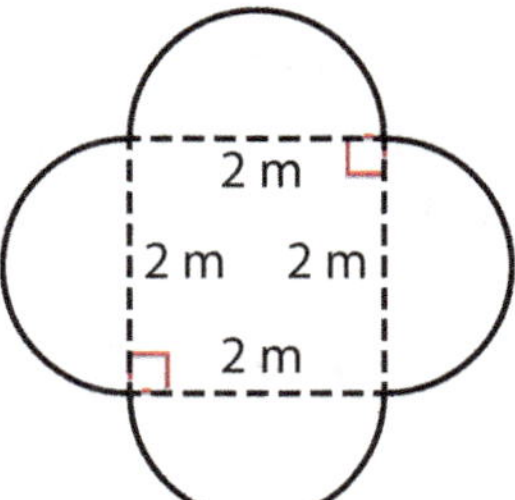

4. arbelos design using 3 semicircles
   $0.5\pi(3^2) - 2[0.5\pi(1.5)^2] = 2.25\pi \approx 7.065$ in.^{2}

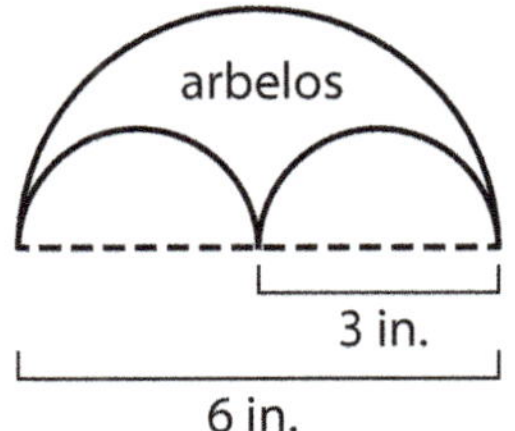

Find the volumes of the given derived figures as well as other parameters.

5. The sides of a concrete vessel at a treatment plant are constructed as a 12 ft square prism that is 16 ft high with a centered cylindrical hole that is 10 ft in diameter and 16 ft long. How much concrete will it take to construct the walls of this vessel?
   vol. sides = vol. prism − vol. cylinder = $12(12)(16) - \pi(5^2)(16) \approx 1048$ ft^3

6. A cardboard box containing 24 cans of soft drink has inside dimensions of 27 cm × 40 cm × 12 cm. A soft drink can is a circular cylinder with a diameter of 6.5 cm and a height of 12 cm. Find what percent of a full case of soft drinks is empty space.
   vol. of empty space = vol. of prism − 24(vol. of a cylinder) = $27(40)(12) -$
   $24[\pi(3.25^2)(12)] \approx 3408.12$ cm^3; percent empty $= \dfrac{3408.12}{12{,}960} \approx 0.26 = 26\%$

7. A 10 ft length of polyvinyl chloride (PVC) pipe has a 4 in. inside diameter and a wall thickness of 0.25 in. If PVC weighs 0.0506 lb per in.^{3}, use the volume of plastic material to find how much this pipe weighs.   vol. material = $\pi(2.25^2)(120) - \pi(2^2)(120)$
   ≈ 400.35 in.^{3}; weight of material = $400.35(0.0506) \approx 20.25$ lb

8. A barbell is made of 2 cast-iron cylinders, 12 cm in diameter and 8 cm long, joined by a round bar, which is 16 cm long and 2.4 cm in diameter. Using the total volume of the 3 parts and the fact that cast iron weighs 7 g/cm^3, find the weight of the barbell in kilograms. (Remember 1 kg = 1000 g.)
   vol. of material = $2[\pi(6^2)(8)] + \pi(1.2^2)(16) \approx 1881$ cm^3;
   weight of material = $1881(7) = 13{,}167$ g = 13.167 kg

Surface Area and Volume of Cans

(Mixed Practice; use after Section 11.6)

Name _______________________________________

Tim's Tin Cans manufactures tin cans for various companies.

1. A tuna can measures 10 cm across and 5 cm high.

    a. How much tin will Tim need to make the whole can? 314 cm²

    b. How much paper will Tim need for the label? 157 cm²

    c. How many cubic centimeters of tuna will fit into the can? 392.5 cm³

2. A soup can measures 6.5 cm across and 10 cm high.

    a. How much tin will Tim need to make the whole can? 270.4 cm²

    b. How much paper will Tim need for the label? 204.1 cm²

    c. How many cubic centimeters of soup will fit into the can? 331.7 cm³

3. A #10 can of baked beans for a restaurant measures 15.5 cm across and 18 cm high.

    a. How much tin will Tim need to make the whole can? 1253.3 cm²

    b. How much paper will Tim need for the label? 876.1 cm²

    c. How many cubic centimeters of baked beans will fit into the can? How many liters?
    3394.7 cm³, ≈ 3.4 L

How did civilizations keep track of time before there were watches and clocks? Sundials, of course—but they were notoriously unreliable at night and in cloudy weather. Water clocks, developed by the Egyptians over 3000 years ago, became a common way of telling time. An internet search for *clepsydra* will return multiple examples of very simple and very complex water clocks. *Clepsydra* comes from Greek words that mean "stealer of water." Both the Greeks and the Romans also used water clocks to time shorter events, such as arguments in legal proceedings. How would you construct a water timer? This project provides lots of opportunities for you to exercise your problem-solving abilities in concept, design, and construction of a drip water timer.

PROCEDURES

Planning the Design

1. Conduct research on water clocks. One of the problems to solve in designing a drip timer is how to keep the dripping consistent. As a container empties through a small hole, the rate decreases as the water level decreases. Make sure you address this issue in your design.

2. With your group (or individually), draw and label a proposed design for a water timer.

Testing the Design

1. When approved by your teacher, construct the drip timer according to your plan.

2. Check your design for ease of use and accuracy.

Refining the Design

1. Use the results from your test to refine your design and correct any deficiencies.

2. Draw and label a new design diagram.

3. Build a new water timer that incorporates your revised design.

Retesting the Design

1. Use your timer 3 separate times.

2. Record the results of the accuracy of your timer on the 3 trials, as well as observations on how easy it was to use.

3. Submit your original and revised designs along with the time trial results and your observations on the revised timer design.

Chapter 11 Cumulative Review

Name ______________________________

_____C_____ **1.** Simplify $2 + 15 - 3^2$. **[1.7]**

    A. 42   **B.** 11

    C. 8   **D.** none of these

_____A_____ **2.** Simplify $7 - (-10)$. **[2.4]**

    A. 17   **B.** -3

    C. 3   **D.** none of these

_____A_____ **3.** Simplify the following expression: $\frac{x^3 y^4}{x^2 y^7}$. **[3.7]**

    A. $\frac{x}{y^3}$ **B.** xy^2

    C. $\frac{x^5}{y^{11}}$ **D.** none of these

_____A_____ **4.** Which number is not equivalent to the others? $9\frac{3}{4}, \frac{27}{4}, 9\frac{12}{16}, \frac{39}{4}$ **[4.4]**

    A. $\frac{27}{4}$ **B.** $9\frac{12}{16}$

    C. $\frac{39}{4}$ **D.** all are equivalent

_____C_____ **5.** Add: $4\frac{5}{6} + (-1\frac{1}{2})$. **[5.3]**

    A. $2\frac{1}{3}$ **B.** $6\frac{1}{3}$

    C. $3\frac{1}{3}$ **D.** none of these

_____B_____ **6.** Which of the expressions below is equivalent with $\frac{1}{4}x + \frac{5}{8}$? **[5.7]**

    A. $\frac{1}{4}(x + \frac{5}{8})$ **B.** $\frac{1}{4}(x + \frac{5}{2})$

    C. $\frac{1}{4}(x + \frac{3}{8})$ **D.** none of these

_____C_____ **7.** Solve for x: $x + 315 = 275$. **[6.2]**

    A. $x = 40$ **B.** $x = -140$

    C. $x = -40$ **D.** none of these

_____D_____ **8.** Three less than the quotient of 36 and a number is 1. What is the number? **[6.4]**

    A. 3   **B.** 4

    C. -12 **D.** none of these

_____B_____ **9.** Justin worked for 5.5 hr in Mrs. Hill's yard on Saturday. At the end of the day, Mrs. Hill gave Justin $65. Use unit rates to determine Justin's hourly rate. **[7.1]**

    A. $13.00 per hour   **B.** $11.82 per hour

    C. $8.46 per hour   **D.** none of these

_____C_____ **10.** Simplify: $\frac{\frac{2}{5}}{3}$. **[7.2]**

    A. $\frac{15}{2}$ **B.** $1\frac{1}{5}$

    C. $\frac{2}{15}$ **D.** none of these

_____D_____ **11.** Change this percent to a fraction in lowest terms: 275%. **[8.1]**

    A. $2\frac{75}{100}$ **B.** $27\frac{1}{2}$

    C. $\frac{4}{11}$ **D.** none of these

_____ C ___**12.** The number 32 is 40% of what number? **[8.4]**

    A. 12.8   **B.** 0.8

    C. 80   **D.** none of these

_____ C ___**13.** Find the percent change, to the nearest percent, if a $110 item is reduced to $90. **[8.6]**

    A. 22%   **B.** 20%

    C. 18%   **D.** none of these

_____ D ___**14.** To the nearest ton, 874,592 lb is equal to how many tons? **[9.2]**

    A. 475 tons   **B.** 436 tons

    C. 447 tons   **D.** none of these

_____ A ___**15.** The measure 45 L is equivalent to which of the following? **[9.3]**

    A. 4500 cL   **B.** 4500 dL

    C. 0.45 daL   **D.** none of these

_____ C ___**16.** Knowing that there are about 1.61 km in a mile, change 12 mi to kilometers. **[9.4]**

    A. 0.13 km   **B.** 7.45 km

    C. 19.32 km   **D.** none of these

_____ D ___**17.** If 2 angles have a sum of 90°, they are what kind of angles? **[10.2]**

    A. adjacent angles   **B.** vertical angles

    C. supplementary angles   **D.** none of these

_____ C ___**18.** A quadrilateral has sides of 8 in., 2 ft, 1 yd, and 1 ft 3 in. Find the perimeter. **[10.3]**

    A. 82 in.   **B.** 2 yd 1 ft

    C. 6 ft 11 in.   **D.** none of these

_____ A ___**19.** A triangle with angles of 72°, 72°, and 36° and two sides of 10 units is what type of triangle? **[10.4]**

    A. isosceles and acute   **B.** scalene and acute

    C. obtuse and isosceles   **D.** none of these

_____ B ___**20.** Find the area of a trapezoid with bases of 10 and 14 cm and an altitude of 6 cm. **[10.5]**

    A. 144 cm²   **B.** 72 cm²

    C. 84 cm²   **D.** none of these

12 Probability

Simple Events

Name ________________________

(Extra Practice; use after Section 12.1)

Mr. Jones filled his homemade "vending machine" with the candies listed below for the students to receive prizes for daily math competitions. Winners push a button 1 time and receive a random candy.

36	Skittles® candies	18	Reese's Pieces® candies
72	Starburst® candies	120	Dum Dum® lollipops
60	Sour Patch Kids® candies	12	Smarties® candies
42	M&M's® candies		

Calculate the probability (in fraction and percent form) that the first winner of the day will get each of the available treats.

$\frac{1}{10}$, 10% **1.** Skittles candies

$\frac{1}{5}$, 20% **2.** Starburst candies

$\frac{1}{6}$, 16.7% **3.** Sour Patch Kids candies

$\frac{7}{60}$, 11.7% **4.** M&M's candies

$\frac{1}{20}$, 5% **5.** Reese's Pieces candies

$\frac{1}{3}$, 33.3% **6.** Dum Dum lollipop

$\frac{1}{30}$, 3.3% **7.** Smarties candies

$\frac{3}{10}$, 30% **8.** Find the probability that a student will receive Skittles candies or Starburst candies.

$\frac{5}{6}$, 83.3% **9.** Find the probability that a student will receive a candy that is not Reese's candies or M&M's candies.

Experiment Day!

(Mixed Practice; use after Section 12.3)

This activity will work best with students in groups. One option would be to allow each group to repeatedly perform a different experiment in a set period of time. Choose 1 student to be the recorder while the other group members take turns repeating the experiment.

Another option would be to rotate groups of students to different experiment stations. Each experiment could be repeated several times so all group members can participate.

Some ideas for fair experiments:

1. Flip a coin. Record heads or tails.

2. Pull marbles or colored balls from a bag. Choose a marble, record the color, put it back, and repeat. There could be multiple stations with this theme with different numbers of marbles.

3. Select tiles from a game like Scrabble or Bananagrams. Choose a tile, record the letter, put it back, and repeat. A simpler alternative might be to record if the letter is a consonant (*C*) or a vowel (*V*).

4. Pick a card from a deck of cards. Record the color or suit of the card, return it, and repeat.

5. Use or make a spinner. If you do not have a commercial one, you can divide a circle into segments, then use a pencil and a large paper clip as the spinner. You can make different ones with different numbers/sizes of slices.

6. Roll a number cube and record the number that comes up.

Student Steps:

1. Calculate the theoretical probability for every possible outcome of each experiment you perform. Record answers as both fractions and percents.

2. Choose a team member to record the results. A different team member should be used to record results for each additional experiment performed.

3. Perform the experiment(s) as many times as possible in the time allotted.

4. Calculate the experimental probability of all of the outcomes in your experiment. If multiple experiments were performed, calculate the experimental probability for the final experiment only.

5. Compare the theoretical probability with the experimental probability.

Record your calculations and your findings here. Use as many columns as needed.

Experiment Name:

Result Name	Theoretical Probability	Result Frequency (use tally marks to record)						Experimental Probability

Compound Events

(Extra Practice; use after Section 12.4)

Name ________________________________

Draw a tree diagram for each of the following exercises and use it to calculate the answer.

_____24_____ **1.** A woman has 3 black dresses that are appropriate for a special occasion; call them b_1, b_2, and b_3. She can wear red, black, gray, or white shoes and has 2 pieces of jewelry to choose between, a string of pearls or a ruby brooch. Draw a tree diagram to show how many different ways she can choose among these items.

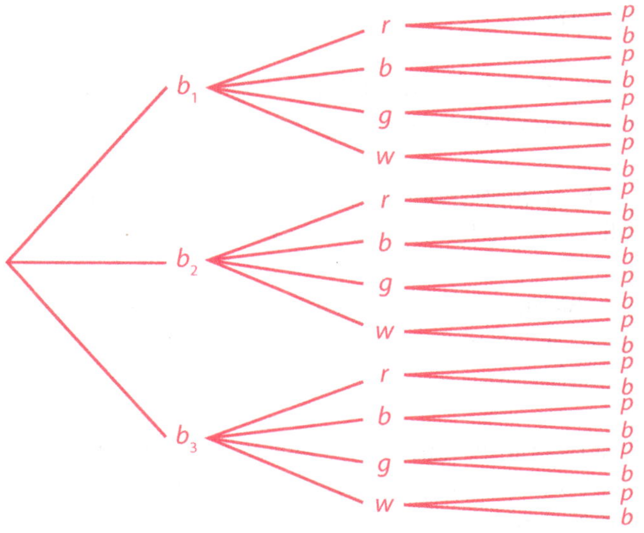

_____36_____ **2.** Find the number of possible ways to build a sandwich if only 1 meat, 1 spread, 1 kind of lettuce, and 1 type of pickle is used. There are 3 meats to choose from (roast beef, ham, and turkey), 3 spreads to put on it (mayo, mustard, and spicy mustard), 2 kinds of lettuces (head and leaf), and 2 kinds of pickles (sweet and dill).

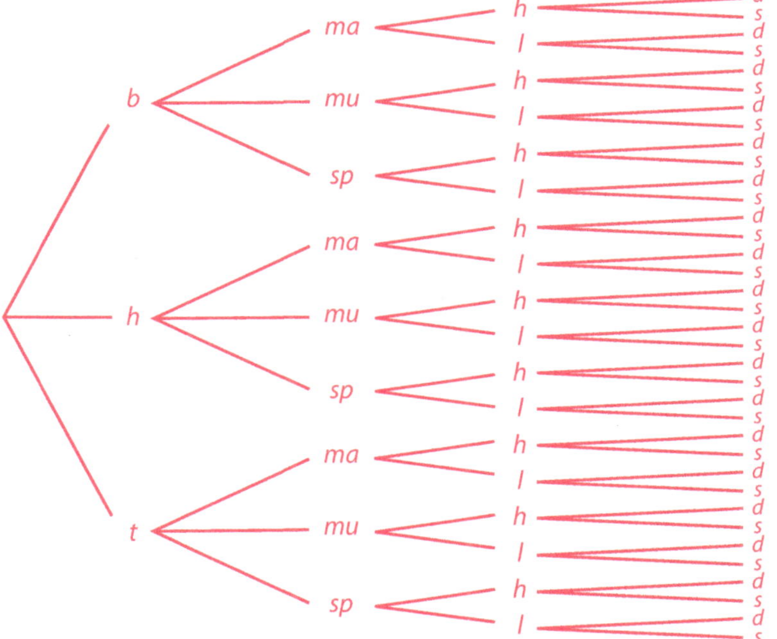

__1000__________________ **3.** In 1921, Alaska issued its first license plates, consisting simply of 3 numeric digits.

__10,000_________________ **4.** By 1925, Alaska had added a fourth digit to their plates.

__17,576,000______________ **5.** Many states, like Minnesota, Kansas, and South Carolina, typically contain 3 alphabetic characters followed by 3 numeric digits.

__67,600,000______________ **6.** A more populous state, like Illinois, might have 2 alphabetic characters followed by 5 numeric digits.

__19,656,000______________ **7.** Suppose both alphabetic characters and all 5 numeric digits on an Illinois plate have to be different.

__30,240_________________ **8.** Suppose the alphabetic characters on all government-owned cars in Illinois have to be *GO*, and numeric digits must be different.

__175,760,000_____________ **9.** The most populous states, like California, have 3 alphabetic characters and 4 numeric digits. Assume the letters/digits can be repeated.

Counting at the Park

(Mixed Practice; use after Section 12.4)

Jamin and Jasmine Jantz are excited because their family is going to an amusement park for the day next Saturday. Mr. Jantz has already bought their "passbooks." Jamin is most interested in the food choices, which are displayed in the following table:

Choose a Meal	Choose a Side	Choose a Drink
cheeseburger	fries	freestyle soda
footlong hot dog	macaroni and cheese	bottled water
chicken tenders		
BBQ		
pizza		

1. Make a tree diagram or an organized list of possible meal combinations if each combination includes a meal, a side, and a drink.

ChB/Fr/FS	HD/Fr/FS	CT/Fr/FS	BBQ/Fr/FS	Pzz/Fr/FS
ChB/Fr/H2O	HD/Fr/H2O	CT/Fr/H2O	BBQ/Fr/H2O	Pzz/Fr/H2O
ChB/MC/FS	HD/MC/FS	CT/MC/FS	BBQ/MC/FS	Pzz/MC/FS
ChB/MC/H2O	HD/MC/H2O	CT/MC/H2O	BBQ/MC/H2O	Pzz/MC/H2O

Additional options for exercise 4.

ChB/CS/FS	HD/CS/FS	CT/CS/FS	BBQ/CS/FS	Pzz/CS/FS
ChB/CS/H2O	HD/CS/H2O	CT/CS/H2O	BBQ/CS/H_2O	Pzz/CS/H_2O

__20__

2. Determine the total number of combinations.

__$5 \times 2 \times 2 = 20$__

3. Use the Fundamental Counting Principle to verify your answer.

__$5 \times 3 \times 2 = 30$__

4. Find the total number of combinations if cole slaw is included as a choice of side.

An amusement park has 3 categories of rides. Jasmine's passbook dictates that she can ride 3 rides from category A, 4 rides from category B, and 3 rides from category C. Jasmine decides to begin the day with category A rides and end with category C rides.

Category A Kids and Family Rides	Category B Moderate Thrill Rides	Category C Ultimate Thrill Rides
1. Back Porch Swings	1. Bermuda Triangle ●	1. Danger Zone ●
2. Bumper Cars	2. Ferris Wheel	2. Niagara Falls ●
3. Carousel	3. Haunted House	3. Lightning Strike ●
4. Jungle Cruise ●	4. Log Flume ●	4. Rocket Launch
5. Locomotive ♇	5. Pirate Ship	5. Slingshot
6. Kiddy Hawk Flyers ♇	6. River Raft Race ●	6. Viper ●
7. Mini Mine Train ●	7. Runaway Mine Train ●	
8. Model Rocket ♇	8. Scrambler	
9. Treehouse ♇		

Jamin used the idea of the license plate problems (See p. 479 of the Student Text) and set up the following template to aid in the solving process:

$$\underline{\quad}\ \underline{\quad}\ \underline{\quad} \times \underline{\quad}\ \underline{\quad}\ \underline{\quad}\ \underline{\quad} \times \underline{\quad}\ \underline{\quad}\ \underline{\quad} = \underline{\qquad}$$

| Category A | Category B | Category C | Total |

Fill in the blanks with the number of choices in each category. Multiply the numbers together (Fundamental Counting Principle) to determine the total number of choices.

5. Determine the number of choices if a ride can be ridden only 1 time.

$$\underline{9} \times \underline{8} \times \underline{7} \times \underline{8} \times \underline{7} \times \underline{6} \times \underline{5} \times \underline{6} \times \underline{5} \times \underline{4} = \underline{101{,}606{,}400}$$

| Category A | Category B | Category C | Total |

6. Determine the number of choices if the first ride in every category is a water ride ● and the last ride of the day is a roller coaster ●.

$$\underline{1} \times \underline{8} \times \underline{7} \times \underline{3} \times \underline{7} \times \underline{6} \times \underline{5} \times \underline{1} \times \underline{5} \times \underline{3} = \underline{529{,}200}$$

| Category A | Category B | Category C | Total |

7. Determine the number of choices if the rides with the ♇ symbols are reserved for shorter guests.

$$\underline{1} \times \underline{4} \times \underline{3} \times \underline{3} \times \underline{7} \times \underline{6} \times \underline{5} \times \underline{1} \times \underline{5} \times \underline{3} = \underline{113{,}400}$$

| Category A | Category B | Category C | Total |

Independent vs. Dependent Events

Name ________________________

Find the probability of the following independent events.

____$\frac{1}{12}$____   **1.** Find the probability of getting heads on a coin flip followed by rolling a 6 on a number cube.

____$\frac{1}{8}$____   **2.** Find the probability of flipping heads on a coin 3 times in a row.

____$\frac{1}{16}$____   **3.** A bag contains 7 red marbles, 4 blue ones, 3 green ones, and 2 white ones. Find the probability of drawing 2 blue marbles in a row, assuming that you replace the marble after the first draw.

____$\frac{7}{64}$____   **4.** Using the same bag as the previous problem, find the probability of drawing a blue marble, replacing it, and then drawing a red marble.

In exercise 3, if a blue marble was pulled out but not replaced, then on the second draw there would be only 15 marbles in the bag. This is a **dependent** event since the second draw is dependent on the result of the first draw. The probability of dependent events is still determined by finding the probability of each event and multiplying them together.

____$\frac{1}{20}$____   **5.** Using the same bag as in problem 3, find the probability of drawing a blue marble, keeping it, and then drawing a green one.

____$\frac{1}{240}$____   **6.** What is the probability of drawing 2 white marbles one after the other followed by a green one (without replacement between draws)?

B   1. Choose the correct letter for the 2 values between which $\sqrt{94}$ falls. **[1.6]**

    A. 9.5 and 9.6   **B.** 9.6 and 9.7

    C. 9.4 and 9.5   **D.** none of these

B   2. Simplify $-14 - 6$. **[2.4]**

    A. 20   **B.** -20

    C. -8 **D.** none of these

B   3. Write 234,000,000 in scientific notation. **[3.8]**

    A. 234×10^6 **B.** 2.34×10^8

    C. 23.4×10^7 **D.** none of these

B   4. Which number is irrational? **[4.5]**

    A. $\sqrt{324}$ **B.** $\sqrt{482}$

    C. $\sqrt{361}$ **D.** all are rational

D   5. Add: $6\frac{3}{8} + 13\frac{5}{6}$. **[5.3]**

    A. $20\frac{19}{24}$ **B.** $20\frac{5}{12}$

    C. $19\frac{5}{24}$ **D.** none of these

D   6. Solve for y: $\frac{y}{-4} < 12$. **[6.7]**

    A. $y < -3$ **B.** $y < -48$

    C. $y > -3$ **D.** $y > -48$

B   7. Which ratio is proportional to the ratio $12 : 18$? **[7.3]**

    A. $3 : 2$ **B.** $24 : 36$

    C. $9 : 6$ **D.** all are proportional

D   8. Find 25.5% of 82 rounded to the nearest tenth. **[8.3]**

    A. 20.0   **B.** 20.8

    C. 21.5   **D.** none of these

C   9. If a coat sold for $62 when discounted 35%, what was the original cost? **[8.7]**

    A. $109.45   **B.** $177.14

    C. $95.38   **D.** none of these

A   10. Convert 13 km to meters. **[9.3]**

    A. 13,000 m   **B.** 130 m

    C. 1300 m   **D.** none of these

A   11. Knowing that there are about 1.06 quarts in a liter, change 8 L to gallons. **[9.4]**

    A. 2.12 gal   **B.** 33.92 gal

    C. 1.89 gal   **D.** none of these

_____B_____ **12.** A bird needs to eat about half of its weight every day. So how many pounds of food would a 3 oz robin need in a year? **[9.5]**

    A. 68.48 lbs   **B.** 34.22 lbs

    C. 15.21 lbs   **D.** 1900 lbs

_____C_____ **13.** When 2 lines intersect, what do they always form? **[10.2]**

    A. at least 2 right angles   **B.** 2 complementary angles

    C. 2 pairs of vertical angles   **D.** none of these

_____A_____ **14.** Find the area of a rectangle with a base of 2 ft and a height of 8 in. **[10.4]**

    A. 192 in.^{2} **B.** 0.21 yd^2

    C. 1.25 ft^2 **D.** none of these

_____C_____ **15.** Find the area of a trapezoid with 1 side length of 8 ft which is perpendicular to the bases of 14 ft and 11 ft. **[10.5]**

    A. 106.25 ft^2 **B.** 200 ft^2

    C. 100 ft^2 **D.** none of these

_____B_____ **16.** If 2 sides of a right triangle are 9 ft and 15 ft, find the hypotenuse (to the nearest tenth). **[10.6]**

    A. 12.0 ft   **B.** 17.5 ft

    C. 17.4 ft   **D.** none of these

_____B_____ **17.** Find the circumference (to the tenth) of a circle with a diameter of 9 in. **[11.1]**

    A. 56.5 in.   **B.** 28.3 in.

    C. 63.6 in.   **D.** none of these

_____C_____ **18.** If $\triangle ABC \sim \triangle RSF$, then **[11.2]**

    A. $AB = RS.$ **B.** $\overline{BC} \cong \overline{SF}.$

    C. $\angle A \cong \angle R.$ **D.** none of these.

_____B_____ **19.** Find the surface area of a rectangular prism with $l = 9$ ft, $w = 3$ ft, and $h = 7$ ft. **[11.3]**

    A. 189 ft^2 **B.** 222 ft^2

    C. 111 ft^2 **D.** none of these

_____A_____ **20.** Find the volume of a triangular prism with bases that are right triangles with sides of 3, 4, and 5 units, respectively, and with a height of 8 units. **[11.5]**

    A. 48 units^3 **B.** 108 units^3

    C. 96 units^3 **D.** none of these

Survey Project
(Enrichment; use after Section 13.1)

Name ___________________________________

Example: Fifty people were surveyed with the following responses. Find the ratio of responses to the total surveyed and the percentages.

What is Your Favorite Color?	Number of Responses	Ratio of Responses to the Total Number Surveyed	Percentage
blue	19	$\frac{19}{50}$	38%
purple	11	$\frac{11}{50}$	22%
red	10	$\frac{10}{50}$	20%
green	6	$\frac{6}{50}$	12%
yellow	4	$\frac{4}{50}$	8%
Total	50	$\frac{50}{50}$	100%

Survey at least 50 people and ask them the following questions. Indicate each response with a tally mark. Then, complete the table with the desired information.

1.

What is Your Favorite Soft Drink?	Number of Responses	Ratio of Responses to the Total Number Surveyed	Percentage
Coca-Cola®			
Mountain Dew®			
Dr. Pepper®			
Pepsi®			
7-Up®			
other			
Total			

2.

What is Your Favorite Sport?	Number of Responses	Ratio of Responses to the Total Number Surveyed	Percentage
baseball			
basketball			
football			
soccer			
swimming			
tennis			
other			
Total			

Range, Mode, Median, Mean

Find the range, mode, median, and mean for the given information. Round to the nearest tenth if necessary.

1. Sandi's test scores: 92, 90, 95, 89, 93   range: 6; mode: none; median: 92; mean: 91.8

2. number of phone calls received each day for a week: 30, 45, 36, 30, 37   range: 15; mode: 30; median: 36; mean: 35.6

3. the week's high temperatures: 96°, 98°, 95°, 93°, 96°, 100°, 101°   range: 8°; mode: 96°; median: 96°; mean: 97°

4. number of points scored this month by the Colonial Christian Academy's basketball team: 78, 71, 85, 95, 78, 91   range: 24; mode: 78; median: 81.5; mean: 83

5. Tom's bowling scores: 93, 114, 100, 123, 95, 117   range: 30; mode: none; median: 107; mean: 107

6. test scores for a seventh-grade class: 92, 84, 93, 97, 72, 100, 95, 68, 79, 88, 92, 95, 92, 90, 79, 97, 80, 88, 80, 79, 72, 86, 88, 77, 80, 86, 80, 86   range: 32; mode: 80; median: 86; mean: 85.5

Name ______________________________

This project allows you to use what you are learning about statistics and presentation of data to identify and address a need in an area of your interest. Sometimes our assumptions about what other people need or want can be very misguided. Work hard to develop a survey that will identify true needs.

PROCEDURES

Planning the Design

1. Conduct research or brainstorm ideas to determine a population you are interested in helping in some way. Here are some examples:

   a. residents in a local nursing home

   b. kindergarten students in a local low-income school district

   c. missionary families in a particular geographic area

2. With your group (or individually) develop a survey to explore how you can best help your selected population.

3. Research the best way to administer your survey (a Key Word Search for free survey tools will provide several digital options.)

Testing the Design

1. When approved by your teacher, choose an unbiased sample and administer the survey.

2. Compile survey results.

Refining the Design

1. Use the results from your survey to refine the design of your project.

2. Develop a presentation that uses charts or graphs of data you collected to explain your project.

Retesting the Design

1. Use your presentation to gain support for your project.

2. Conduct the project.

3. Submit your original and revised plans, with survey results and observations on how effective you think your project was in meeting your objectives.

Circle Graphs

Circle graphs show the relationship of parts to the whole and to each other.

The circle graph at right represents Tina's spending for the past month. Observe that Tina has no money budgeted for savings.

Example: Use the circle graph to decide how much Tina spends on rent and utilities.
Answer: 37% of 1900 = 0.37 × 1900 = $703
Tina spends $703 on rent and utilities.

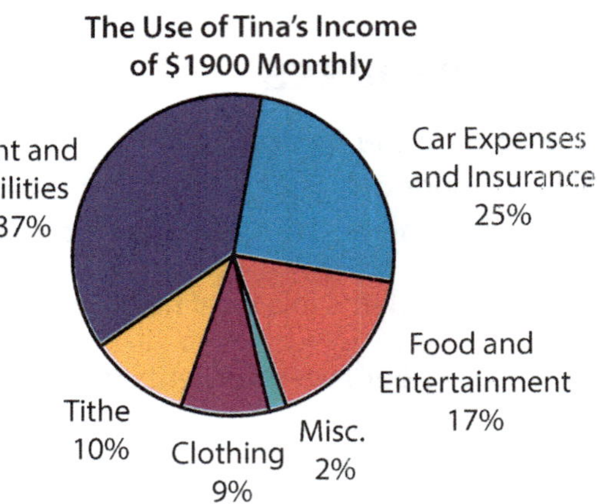

Use the circle graph of Tina's budget to determine the amount spent on

_____$475_____ **1.** car expenses and insurance.

_____$171_____ **2.** clothing.

_____$323_____ **3.** food and entertainment.

_____$38_____ **4.** miscellaneous.

To save $351.50 per month on her rent and utilities, Tina decided to share an apartment with a friend. This enabled her to increase her giving to her church to 12% and to save 8%. She plans to use the remaining amount for recreation.

5. Make a circle graph for Tina's new budget.

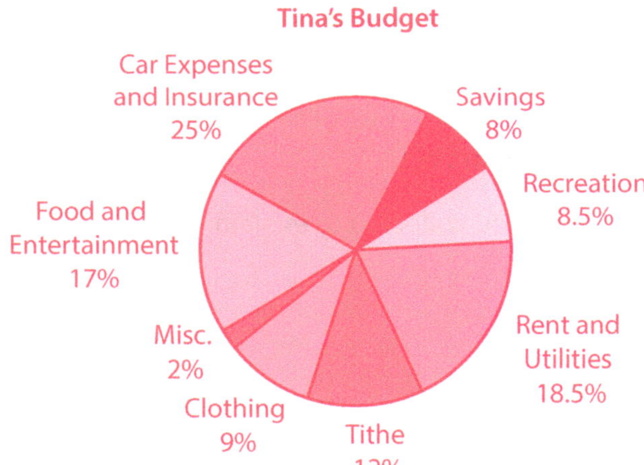

Use the information given and the new circle graph to determine the amount spent on

_____$228_____ **6.** tithes.

_____$152_____ **7.** savings.

_____$475_____ **8.** car expenses and insurance.

_____$161.50_____ **9.** recreation.

_____$2000_____ **10.** What would Tina's income need to be for her 12% "tithe" amount to be $240?

Using Statistics to Persuade

(Problem Solving; use after Section 13.5)

Name _______________________________

1. When an automobile advertisement for Fine Motor Works says that a certain model gets 25 mi/gal on the average, it could be using any one of the measures of central tendency—mean, median, or mode. Given the fuel economy ratings below for 15 of the company's vehicles, find the mean, median, and mode. Which of the measures do you think the company will use to claim that their car is economical?

   fuel economy ratings: 35, 32, 30, 25, 25, 25, 24, 23, 22, 22, 20, 20, 19, 18, 18
   mean: 23.87; median: 23; mode: 25; They will use the mode.

2. Find the mean absolute deviation of Fine Motor Works' fuel economy ratings. Then find the mean and MAD for American Muscle's models and write a sentence comparing the 2 auto makers.

   American Muscle fuel economy ratings: 28, 27, 25, 25, 24, 23, 23, 21, 20, 19, 18
   MAD for FMW: 3.86; mean for AM: 23; MAD for AM: 2.55; Both companies have similar mean fuel economy ratings, but American Muscle's ratings are more consistent.

3. A certain corporation that was trying to convince others how well they compensate their employees stated that the mean salary for their employees was $80,000. However, a check of company records shows that the median salary was $40,000. What might cause such a discrepancy? A few highly compensated executives would raise the mean significantly but not affect the median. For example, 48 employees that each make $40,000 plus 2 executives each making $1,040,000 would give a mean salary of $80,000, but the median salary would still be $40,000.

4. Compare the data in the table to the bar graph to determine if the graph gives a true representation of which school had the best participation in its service clubs. What might cause the bar graph to be misleading? What other information would it be helpful to know? The contracted vertical scale makes it look like Southside High School (SHS) has significantly fewer students in service clubs. It would be helpful to know the number of students in each school.

Number of Students in Service Clubs				
School	EHS	WHS	NHS	SHS
women	49	37	38	32
men	13	21	18	16
Total	62	58	56	48

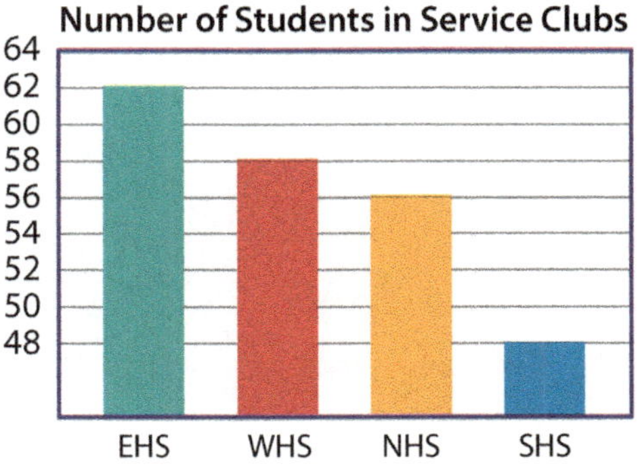

5. Given the following number of students in each school, make a new bar graph based on the percent of students in the schools that are in the service clubs.

Eastside (EHS): 840; Westside (WHS): 920; Northside (NHS): 700; Southside (SHS): 540
Percent of students: EHS 7.4%; WHS 6.3%; NHS 8.0%; SHS 8.9%

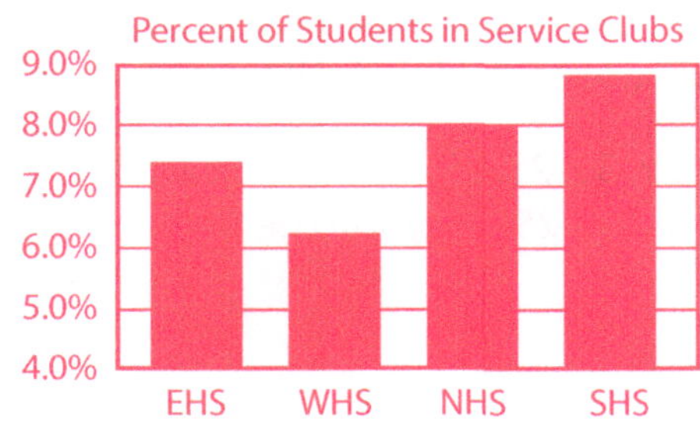

6. Benjamin averaged 68 mi/hr, and it took 1 hr 15 min to drive from his college to his home. On the way back to college, he averaged only 50 mi/hr. He claims that he averaged 59 mi/hr for the round trip. Explain why this is a false average. Find his actual average speed. The claim is false, since he averaged his 2 average speeds. He took longer coming back at a slower rate, so the answer cannot be the average of 68 and 50.

$$\text{mean} = \frac{\text{total distance}}{\text{total time}} = \frac{85+85}{1.25+1.7} = \frac{170}{2.95} = 57.63 \text{ mi/hr}$$

Bar and Line Graphs
(Extra Practice; use after Section 13.5)

Name ______________________________

Bar and line graphs are types of graphs used for specific purposes. Use the bar graph to compare statistics and the line graph to show how something changes over a period of time. You can also use double-bar graphs and double-line graphs to compare sets of data.

Use the bar graph to compare statistics in questions 1–4.

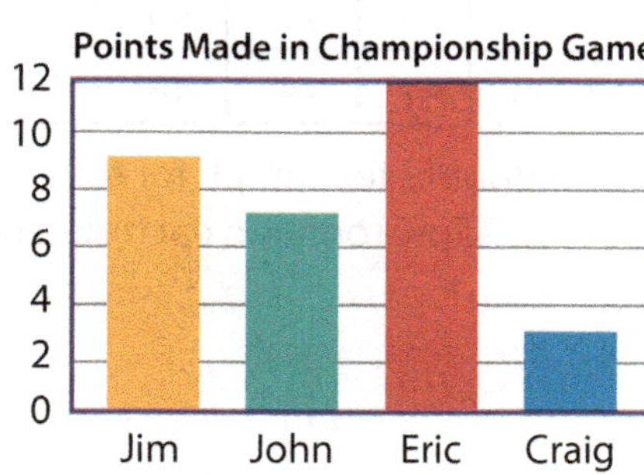

Eric **1.** Who made the most points?

_3_ **2.** If 6 of Jim's points were field goals, how many free throws did he make?

John **3.** Who made 7 points?

9 points **4.** What is the difference between the highest number and the lowest number of points made?

Use the line graph to answer questions 5–7.

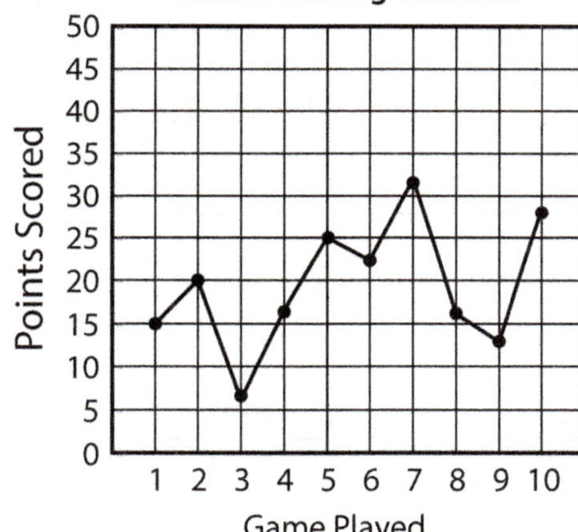

_7_ **5.** In what game did Keith score over 30 points?

4 and 8 **6.** In what games did Keith score the same number of points?

_3; 6_ **7.** In what game did Keith score the lowest number of points? How many did he score?

8. Make a double-bar graph to illustrate the data given below.

Points scored in 2018: Game 1—79 points; Game 2—75 points; Game 3—68 points; Game 4—95 points; Game 5—80 points

Points scored in 2019: Game 1—68 points; Game 2—78 points; Game 3—82 points; Game 4—59 points; Game 5—80 points

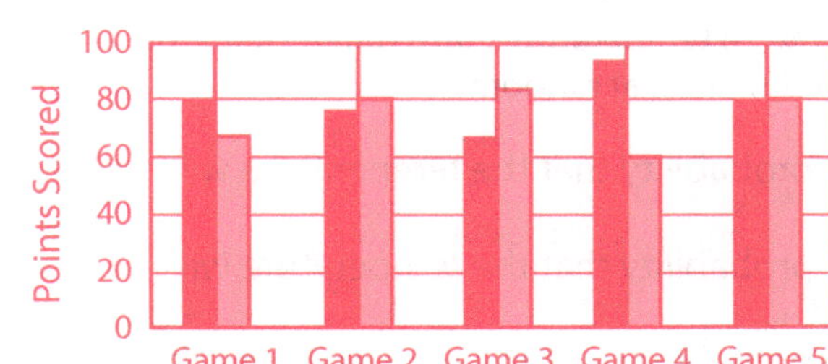

Statistics with Graphs and Diagrams
(Mixed Practice; use after Section 13.5)

Mean Monthly Temperatures (Fahrenheit)												
City	Jan.	Feb.	Mar.	Apr.	May	June	July	Aug.	Sept.	Oct.	Nov.	Dec.
Norfolk, VA	39°	41°	49°	58°	66°	73°	78°	76°	71°	61°	52°	44°
Portland, OR	41°	44°	48°	52°	58°	63°	68°	69°	64°	56°	47°	41°

1. Make a double-line graph for the mean monthly temperatures of Norfolk and Portland. Allow room on each side of the graph to add a box-and-whisker plot later.

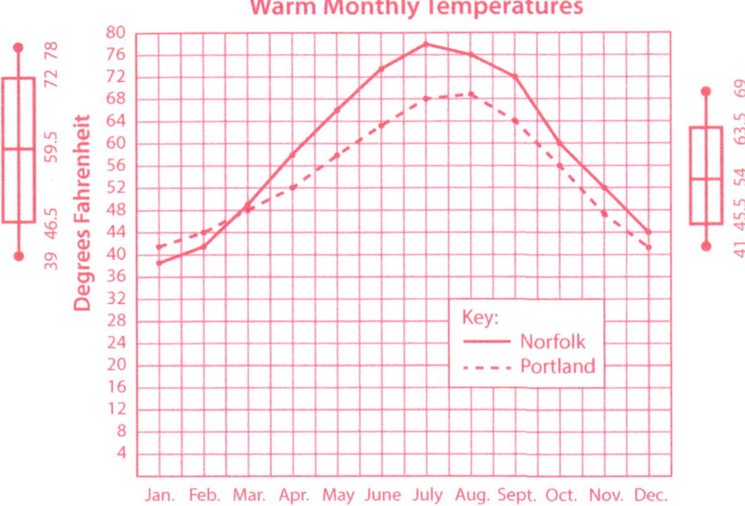

Norfolk: 59°F; Portland: 54.25°F

2. Find the mean of the monthly temperatures for each city.

Norfolk: 59.5°F; Portland: 54°F

3. Find the median of the monthly temperatures for each city.

Norfolk: 39°F

4. Which city has the widest range of temperatures? What is a possible explanation for the difference? Norfolk: 39°F; Portland: 28°F; explanations will vary. Possible answer: The Atlantic temperatures vary more than the Pacific.

lower: 46.5°F; upper: 72°F

5. Find the lower and upper quartiles of the monthly temperatures for Norfolk.

lower: 45.5°F; upper: 63.5°F

6. Find the lower and upper quartiles of the monthly temperatures for Portland.

7. Construct a vertical box-and-whisker plot for Norfolk on the left side of the line graph drawn in exercise 1. Construct a vertical box-and-whisker plot for Portland on the right side of the line graph drawn in exercise 1. Make sure you align the scale of the box-and-whisker plots with the vertical scale of the line graph.

Suppose you write each of the 24 temperatures on a different marble and place them in a bag. Choose 1 marble at random.

$\frac{1}{8}$ **8.** What is the probability that the temperature will read 41°?

$\frac{1}{12}$ **9.** What is the probability that the temperature will read 52°?

$\frac{1}{6}$ **10.** What is the probability that the temperature will read more than 70°?

Name ____

_____A_____ **1.** Estimate the product of 87.41×32.9. **[1.3]**

    A. 2700   **B.** 27

    C. 270   **D.** none of these

_____B_____ **2.** Which is the greatest integer in the following set? $\{0, 1, -2, -150\}$ **[2.2]**

    A. 0   **B.** 1

    C. -2 **D.** -150

_____A_____ **3.** Find the missing multiplicand of $9.23 \times 7.1 = \underline{\hspace{1cm}} \times 9.23$. **[3.2]**

    A. 7.1   **B.** 65.533

    C. 9.23   **D.** none of these

_____B_____ **4.** Find the LCM of 6, 8, and 16. **[4.3]**

    A. 24   **B.** 48

    C. 32   **D.** none of these

_____B_____ **5.** Add $\frac{2}{7} + \frac{7}{15} + \frac{3}{14}$. **[5.1]**

    A. $\frac{14}{29}$ **B.** $\frac{203}{210}$

    C. $\frac{101}{105}$ **D.** none of these

_____C_____ **6.** Solve for z: $\frac{3}{4}(2z + 6) = 18$. **[6.5]**

    A. $z = 3$ **B.** $z = 8$

    C. $z = 9$ **D.** none of these

_____B_____ **7.** Timmy is training his turtle, Skippy, for a race. His latest time on the 9 yd course was $7\frac{1}{5}$ sec, Skippy's personal best. Convert Skippy's rate to a unit rate. **[7.2]**

    A. 65 yd to 1 sec   **B.** $1\frac{1}{4}$ yd to 1 sec

    C. $\frac{4}{5}$ yd to 1 sec   **D.** none of these

_____A_____ **8.** To the nearest percent, what percent of 90 is 17? **[8.4]**

    A. 19%   **B.** 529%

    C. 5%   **D.** none of these

_____C_____ **9.** Shelly's dad is building a rectangular sandbox for her little brother. He plans for it to be 6 ft 3 in. long and 3 ft 5 in. wide. How much wood will he need? **[9.1]**

    A. 18 ft 16 in.   **B.** 18 ft 12 in.

    C. 19 ft 4 in.   **D.** none of these

_____C_____ **10.** Which is the most precise measurement? **[9.3]**

    A. 6.3 m   **B.** 6.3 km

    C. 629 cm   **D.** A and C are the same.

_____B_____ **11.** Which of the following angle measurement/angle classification pairs is incorrect? **[10.1]**

    A. 25°/acute   **B.** 180°/obtuse

    C. 90°/right   **D.** All are correct.

_____B_____ **12.** A quadrilateral with exactly 1 pair of parallel sides is called what? **[10.3]**

    A. parallelogram   **B.** trapezoid

    C. regular quadrilateral   **D.** rhombus

_____A_____ **13.** Find the area of a parallelogram with consecutive sides of 5 ft and 7 ft and with an altitude of 4 ft between the longer sides. **[10.4]**

    A. 28 ft^2 **B.** 96 ft^2

    C. 35 ft^2 **D.** 20 ft^2

_____B_____ **14.** Find the area of a circle with a radius of 2.5 m. Round to the nearest tenth. **[11.1]**

    A. 20.5 m^2 **B.** 19.6 m^2

    C. 21.5 m^2 **D.** none of these

_____C_____ **15.** Two similar figures have corresponding sides of 6 m and 2 m. The larger figure has an area of 30 m^2. Find the area of the smaller figure to the nearest 10th. **[11.2]**

    A. 10.0 m^2 **B.** 5.6 m^2

    C. 3.3 m^2 **D.** not enough information

_____B_____ **16.** Find the surface area of a rectangular prism with dimensions of 9 ft by 12 ft by 15 ft. **[11.5]**

    A. 1620 ft^2 **B.** 846 ft^2

    C. 423 ft^2 **D.** none of these

_____C_____ **17.** A bag contains 6 red marbles, 4 green marbles, 3 yellow marbles, and 1 white marble. If one marble is drawn at random, what is the probability it will be green? **[12.1]**

    A. $\frac{5}{7}$ **B.** $\frac{1}{4}$

    C. $\frac{2}{7}$ **D.** none of these

_____C_____ **18.** What is the probability of rolling an even number on a number cube? **[12.1]**

    A. $\frac{1}{6}$ **B.** $\frac{1}{3}$

    C. $\frac{1}{2}$ **D.** none of these

_____B_____ **19.** If a spinner that is divided into 4 equal segments colored red, yellow, green, and blue is spun 84 times, how many times would you expect it to land on red? **[12.3]**

    A. about 63 times   **B.** about 21 times

    C. about 25 times   **D.** none of these

_____C_____ **20.** If there are 5 ways to do one task, 3 ways to do a 2nd task, and 2 ways to do a 3rd task, how many ways are there to do the 3 tasks in succession? **[12.4]**

    A. 10   **B.** 16

    C. 30   **D.** none of these

Graphy Duck

(Mixed Practice; use after Section 14.1)

Name _______________________

Working down each column, plot each ordered pair, connecting the points with a straight line as you plot them. When you see the word *STOP*, lift your pencil to start a new section.

(–12, –11)	(–9, 2)	(0, 10)	(9, 0)	(1, –7)	(–5, –3)	(–3, 3)	(1, 1)
(–11, –9)	(–10, 2)	(–1, 11)	(9, –3)	(–3, –6)	(0, –1)	(–2, 2)	(1, 2)
(–9, –7)	(–10, 3)	(0, 10)	(7, –5)	(–5, –5)	(4, –3)	(–2, 1)	(2, 3)
(–8, –6)	(–8, 6)	(1, 11)	(4, –6)	**STOP**	(0, –1)	(–3, –1)	(3, 2)
(–9, –7)	(–7, 8)	(0, 10)	(5, –7)	(–5, –3)	(1, –4)	(–4, 0)	(4, 0)
(–11, –7)	(–3, 10)	(2, 9)	(7, –7)	(1, –4)	(0, –6)	(–5, 0)	(2, –1)
(–13, –6)	(–4, 11)	(4, 7)	(8, –9)	(4, –3)	**STOP**	(–4, 0)	(1, 1)
(–15, –4)	(–3, 10)	(6, 3)	(8, –11)	(2, –5)	(–7, –1)	(–3, 1)	(3, 1)
(–15, 0)	(–2, 11)	(5, 2)	**STOP**	(0, –6)	(–5, 0)	(–2, 1)	(4, 0)
(–13, 1)	(–3, 10)	(8, 1)	(4, –6)	(–2, –5)	(–5, 2)	**STOP**	**STOP**

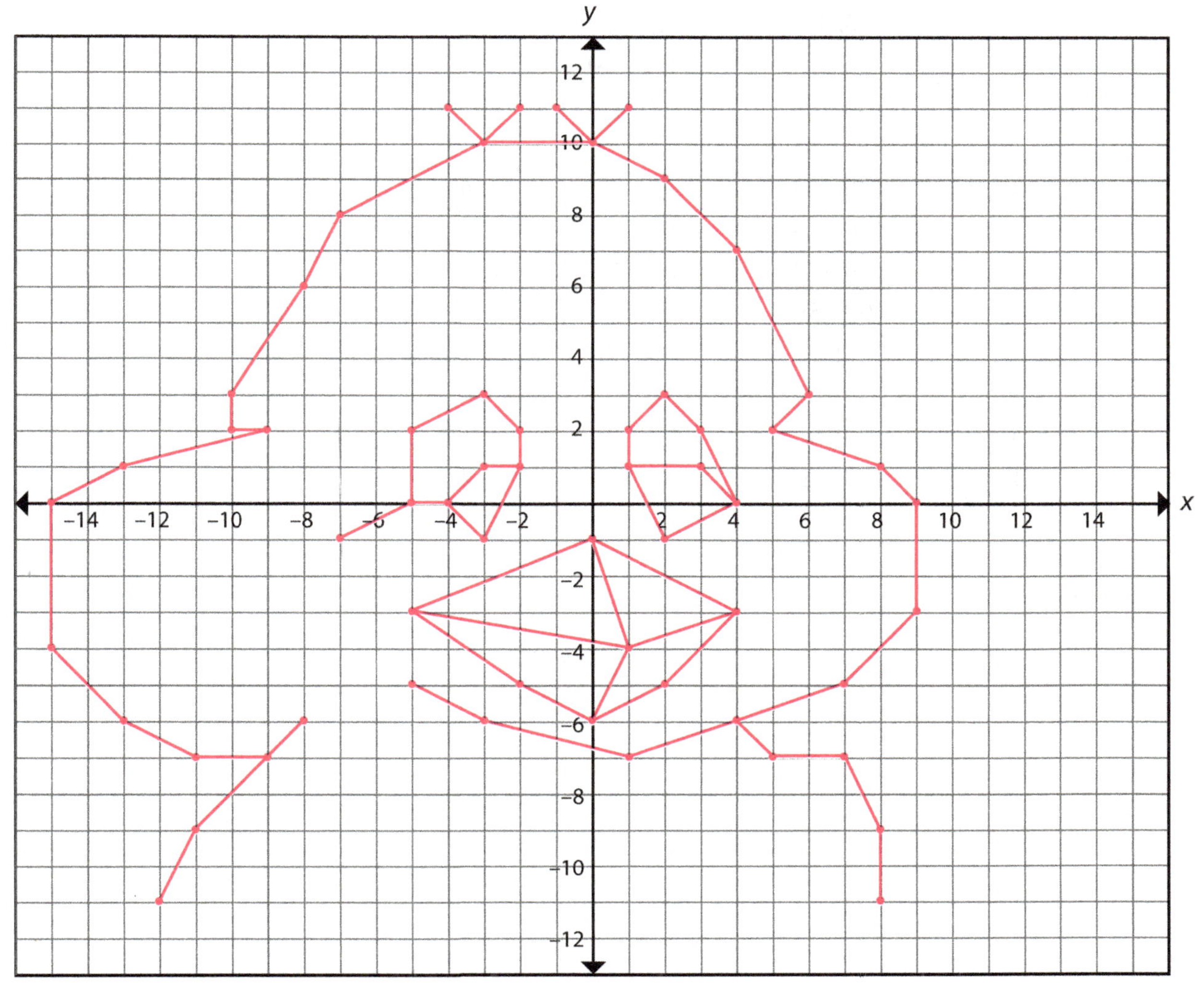

Coordinate Plane

On graph paper draw a pair of axes and plot each ordered pair. Read the hidden message.

(1, 5)	(−3, −3)	(7, 3)	(6, −1)
(−5, −1)	(2, −1)	(−1, 5)	(−8, 1)
(−5, 1)	(6, 1)	(−1, −3)	(−3, 1)
(−5, −3)	(0, 5)	(4, −3)	(−5, 5)
(−2, −1)	(0, 4)	(−5, 3)	(2, −3)
(3, 1)	(−9, 4)	(−2, 2)	(4, −1)
(−8, −2)	(−10, 1)	(6, −4)	(−4, 1)
(−2, 5)	(6, 2)	(5, 5)	(6, −3)
(0, 3)	(−9, −4)	(−9, 3)	(2.5, 2)
(2, 3)	(−5, 2)	(4, −5)	(4, −4)
(−2, −3)	(−5, −5)	(2, −4)	(3.5, 2)
(2, −2)	(5, −3)	(0, −5)	(−5, −4)
(4, 3)	(6, 4)	(6, −5)	(−6, −2)
(0, 2)	(6, 5)	(−3.5, −4)	(4.5, 4)
(−1, 1)	(−7, −3)	(7, 1)	(−9, −5)
(2, −5)	(−5, −2)	(0, 1)	(4, −2)
(−1.5, −2)	(−4, −5)	(7, 5)	(−5, 4)
(6, −2)	(−9, −1)	(−2, 4)	(−0.5, −4)
(3, −1)	(−8, 5)	(6, 3)	(1, −1)
(8, 5)	(−2, 1)	(−9, −3)	(−10, 5)
(1.5, 4)	(−9, 5)	(−2, 3)	(−9, 2)
(−9, 1)	(−9, −2)	(8, 1)	(−2.5, −2)

Identifying Functions
(Extra Practice; use after Section 14.2)

Determine whether each relation is a function.

__yes__ **1.** {(3, 1), (4, 1), (5, 1)}

__no__ **2.** {(−1, 2), (−2, 3), (−1, 3), (−2, 2)}

__yes__ **3.** {(9, 0), (−9, 1), (0, 9), (1, −9)}

__yes__ **4.** {(2, −1), (4, −2), (6, −3), (8, −4)}

__no__ **5.** {(−1, −1), (−1, 2), (−1, 0)}

__yes__ **6.** {(14, 2), (15, 3), (16, 2), (17, 3)}

__yes__ **7.** {(2, 2), (3, 3), (4, 4), (5, 5)}

__no__ **8.** {(−5, 1), (−4, 2), (−3, 3), (−5, 4)}

__no__ **9.** {(1, 1), (1, 2), (1, 3), (1, 4), (1, 5)}

__yes__ **10.** {(7, −2), (6, −2), (5, −2), (4, −2)}

Determine whether each of the following circle mappings represents a function.

__no__ **11.**

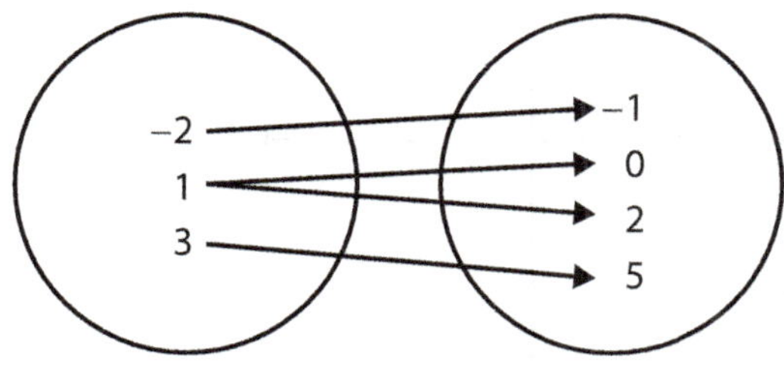

__no__ **12.**

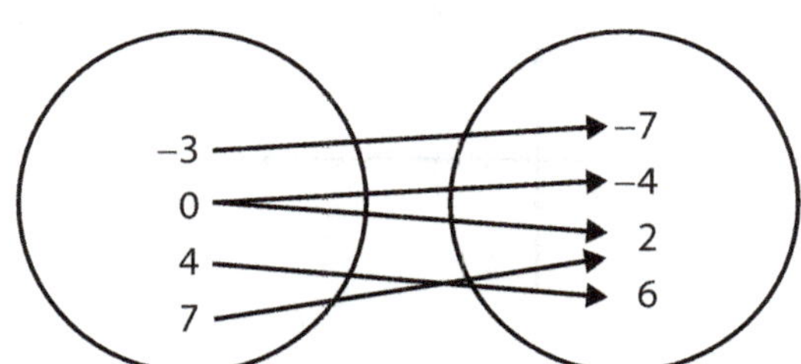

__yes__ **13.**

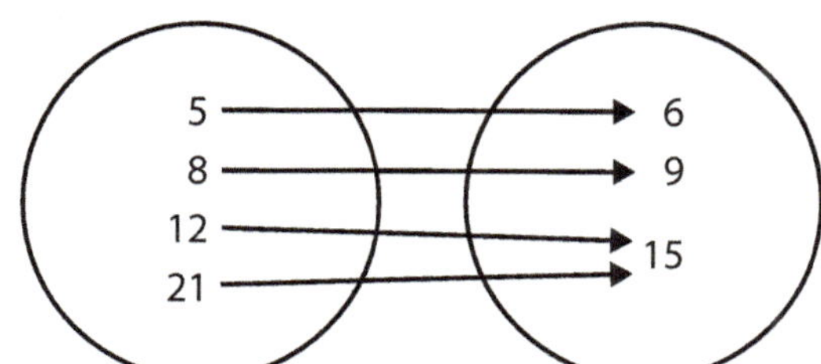

__no__ **14.**

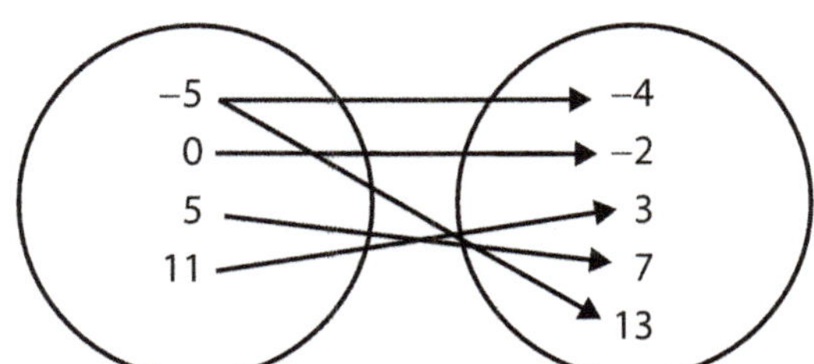

Determine whether the following graphs represent functions.

_____yes__ **15.**

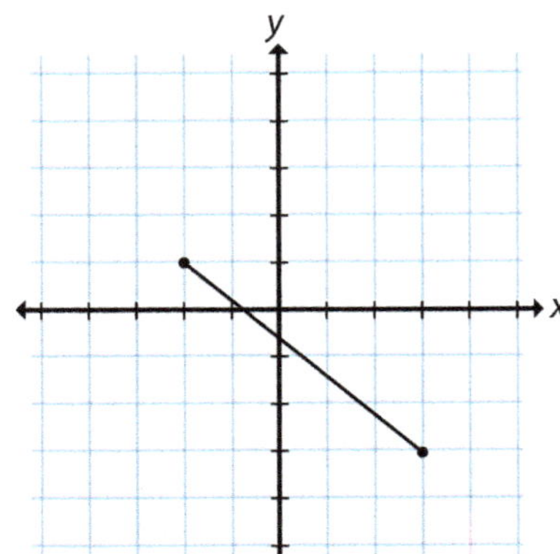

_____no__ **16.**

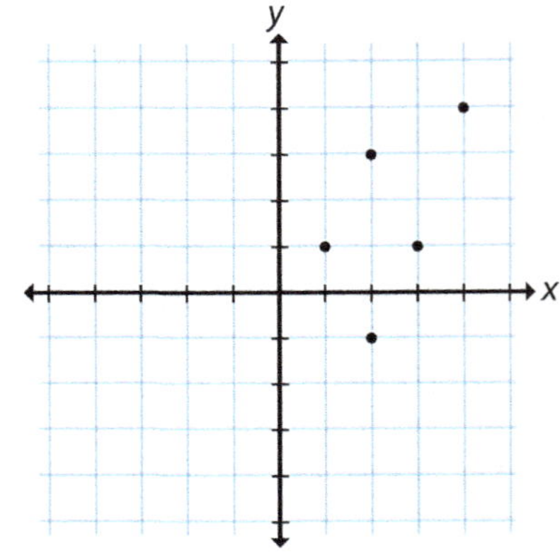

_____no__ **17.**

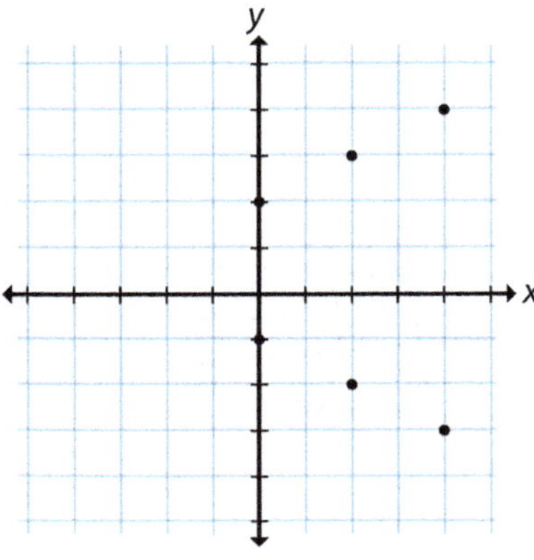

_____yes__ **18.**

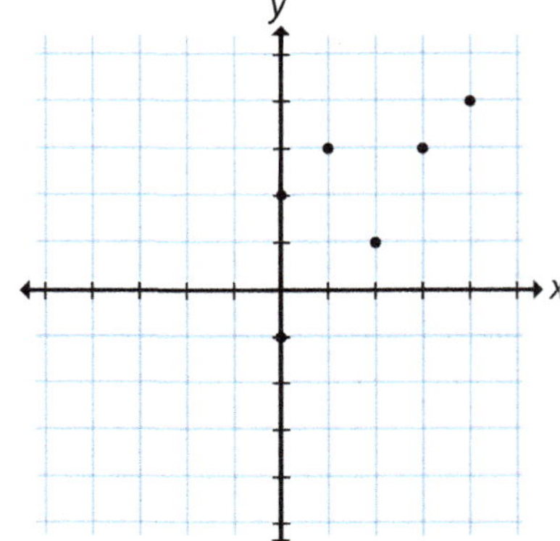

Function Rules

(Extra Practice; use after Section 14.3)

Name _______________________________

Choose the letters of the ordered pairs which satisfy the given function rule. If none of the ordered pairs satisfy the rule, write *none*.

_____**B**_____ **1.** $y = 3x - 4$

       A. $(2, -1)$
       B. $(-2, -10)$
       C. $(9, 22)$

_____**A, C**_____ **2.** $y = |5x - 3|$

       A. $(-7, 38)$
       B. $(-3, 12)$
       C. $(2, 7)$

_____**B**_____ **3.** $y = \frac{x}{4} + 2$

       A. $(64, 14)$
       B. $(-12, -1)$
       C. $(-4, 3)$

_____**C**_____ **4.** $y = \frac{9}{5}x + 32$

       A. $(-10, 16)$
       B. $(35, 85)$
       C. $(-40, -40)$

_____**A**_____ **5.** $y = 15 - 5x$

       A. $(3, 0)$
       B. $(-2, -25)$
       C. $(10, 35)$

_____**none**_____ **6.** $y = 3|2x - 6| - 4$

       A. $(-2, 34)$
       B. $(4, -2)$
       C. $(9, 36)$

_____**A, B, C**_____ **7.** $y = |x| - 2x + 7$

       A. $(2, 5)$
       B. $(-3, 16)$
       C. $(20, -13)$

Find the ordered pairs for each function rule if the *x*-values are −2, 4, and 10.

8. $y = 12x - 3$ $(-2, -27), (4, 45), (10, 117)$

9. $y = \frac{2x - 8}{4}$ $(-2, -3), (4, 0), (10, 3)$

10. $y = |5x - 12|$ $(-2, 22), (4, 8), (10, 38)$

11. $y = \frac{4}{3}x + \frac{5}{3}$ $(-2, -1), (4, 7), (10, 15)$

12. $y = 15 - |7 - 2x|$ $(-2, 4), (4, 14), (10, 2)$

Given that $y = -3(x - 1) + 5$, find the following functional values.

_____5_____**13.** $f(1)$

_____17_____**14.** $f(-3)$

_____−16_____**15.** $f(8)$

_____26_____**16.** $f(-6)$

_____−28_____**17.** $f(12)$

_____35_____**18.** $f(-9)$

Graphing Linear Functions
(Extra Practice; use after Section 14.4)

Name ________________________________

Make a table of values for each function using the given *x*-values; then graph the function on a coordinate plane.

1. $y = 2(x - 3) + 2$; using $x = 0, 2, 3, 4$

x	$2(x - 3) + 2$	y
0	$2(0 - 3) + 2 = 2(-3) + 2 = -6 + 2$	-4
2	$2(2 - 3) + 2 = 2(-1) + 2 = -2 + 2$	0
3	$2(3 - 3) + 2 = 2(0) + 2 = 0 + 2$	2
4	$2(4 - 3) + 2 = 2(1) + 2 = 2 + 2$	4

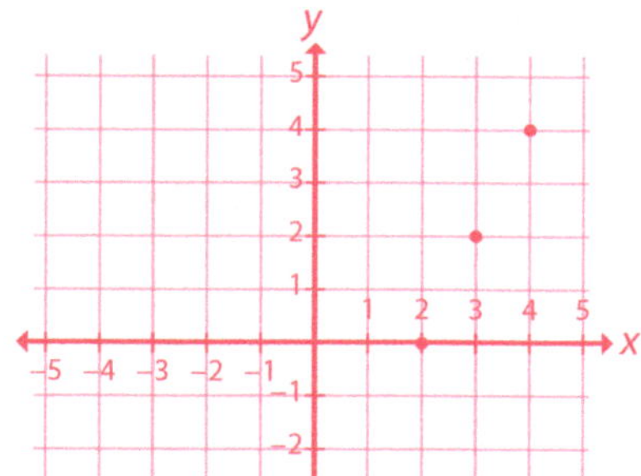

2. $y = -\frac{3}{4}x + 1$; using $x = -4, 0, 2, 4$

x	$-\frac{3}{4}x + 1$	y
-4	$-\frac{3}{4}(-4) + 1 = 3 + 1$	4
0	$-\frac{3}{4}(0) + 1 = 0 + 1$	1
2	$-\frac{3}{4}(2) + 1 = -\frac{3}{2} + 1$	$-\frac{1}{2}$
4	$-\frac{3}{4}(4) + 1 = -3 + 1$	-2

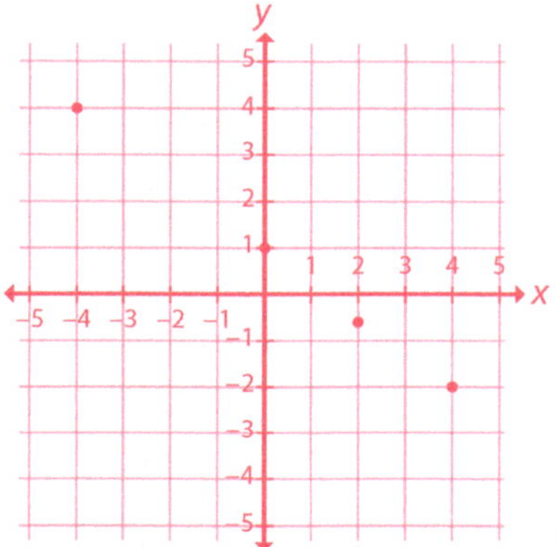

3. $y = 3x$; using $x = -1, 0, 1, 2$

x	$3x$	y
-1	$3(-1)$	-3
0	$3(0)$	0
1	$3(1)$	3
2	$3(2)$	6

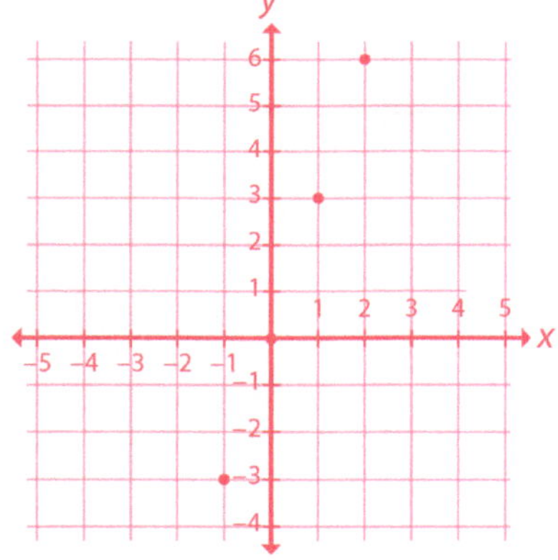

4. $y = 2$; using $x = -2, 0, 1, 3$

x	2	y
-2	2	2
0	2	2
1	2	2
3	2	2

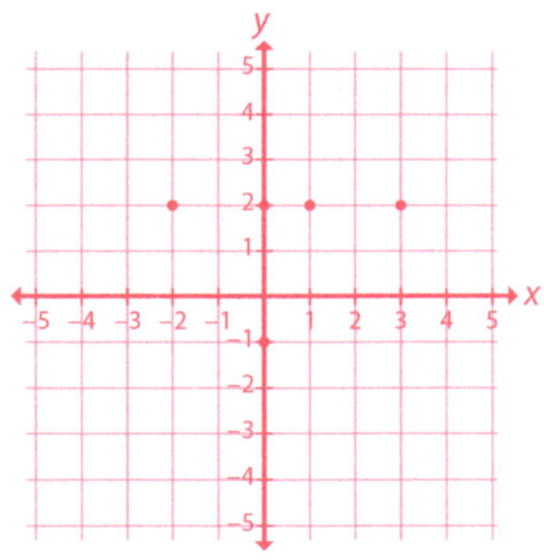

5. $y = -\frac{1}{2}x + 2$; using $x = -2, 0, 4, 6$

x	$-\frac{1}{2}x + 2$	y
-2	$-\frac{1}{2}(-2) + 2 = 1 + 2$	3
0	$-\frac{1}{2}(0) + 2 = 0 + 2$	2
4	$-\frac{1}{2}(4) + 2 = -2 + 2$	0
6	$-\frac{1}{2}(6) + 2 = -3 + 2$	-1

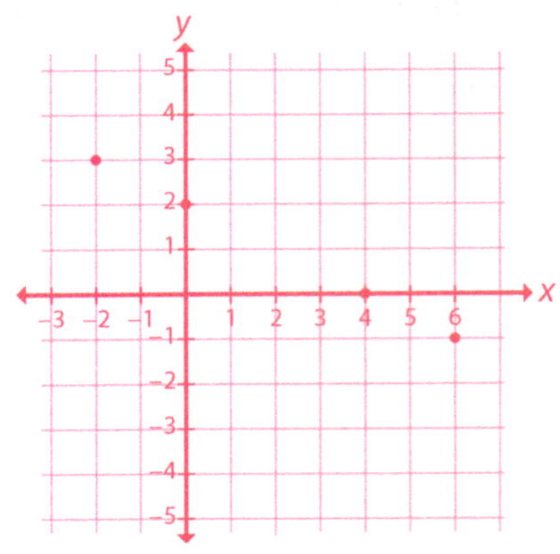

6. $y = 2x - 1$; using $x = -2, -1, 0, 2$

x	$2x - 1$	y
-2	$2(-2) - 1 = -4 - 1$	-5
-1	$2(-1) - 1 = -2 - 1$	-3
0	$2(0) - 1 = 0 - 1$	-1
2	$2(2) - 1 = 4 - 1$	3

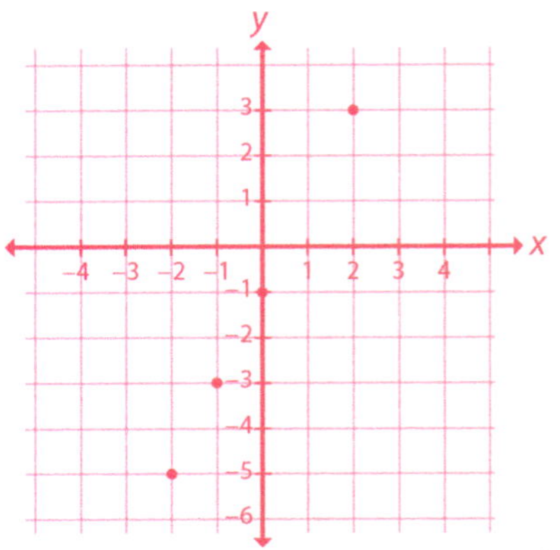

7. $y = \frac{2}{3}(x + 1) - 3$; using $x = -4, -1, 2, 5$

x	$\frac{2}{3}(x + 1) - 3$	y
-4	$\frac{2}{3}(-4 + 1) - 3 = -2 - 3$	-5
-1	$\frac{2}{3}(-1 + 1) - 3 = 0 - 3$	-3
2	$\frac{2}{3}(2 + 1) - 3 = 2 - 3$	-1
5	$\frac{2}{3}(5 + 1) - 3 = 4 - 3$	1

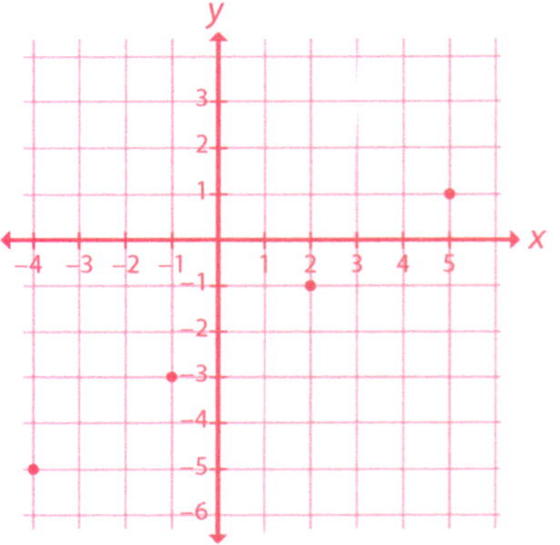

8. $y = |x|$; using $x = -5, -3, 0, 3, 5$

x	$|x|$	y
-5	$|-5|$	5
-3	$|-3|$	3
0	$|0|$	0
3	$|3|$	3
5	$|5|$	5

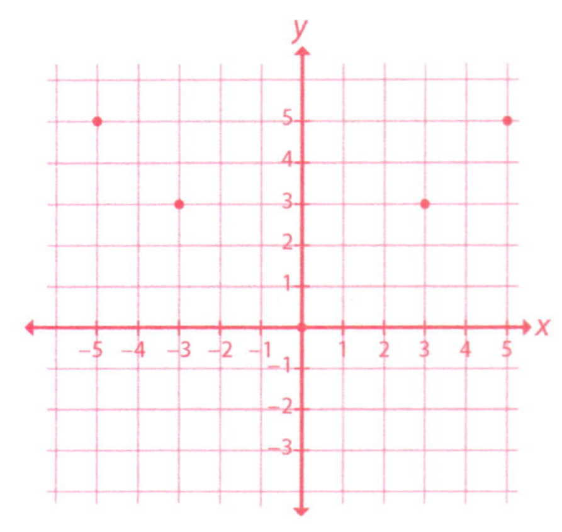

Fundamentals of Math

Graphs from Experimental Ordered Pairs

Name ______________________

The depth of water in a container as a function of the amount of water in the container generates a set of ordered pairs that can be graphed. Plot the amount of water in the container as the *x*-value and the depth as the *y*-value. First, find out how much water the container will hold and divide that amount into 10 or 12 equal increments. Pour the first increment of water into the container and measure the depth; then add the next increment of water and measure that depth. Continue this until the entire container is full. At each step, record in a table the amount of water and the depth.

The measurements are best found using milliliters of water and a ruler that measures in centimeters and millimeters. The depth is hard to read even with see-through containers; you may need to color the water or insert a stick that will show the water line and then measure the wet part of the stick to get the depth. Narrow strips of poster board can also be used to find the successive water depths. Measure the marks on the posterboard strips quickly, because the line will become blurred. Plot the points from the table on graph paper, and then draw a smooth curve connecting the points. Study the example below.

No.	Water in mL	Depth in mm	No.	Water in mL	Depth in mm
1	35	15	6	210	41
2	70	25	7	245	47
3	105	28	8	280	55
4	140	31	9	315	60
5	175	36	10	350	72

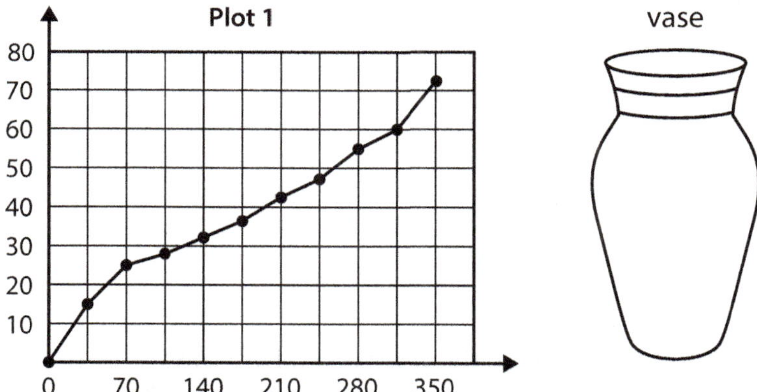

Observe the shape of the graph and note that when the cross section of the container was a small circle the depth changed quickly between increments of water. Likewise, when the cross section was a larger circle, the depth changed less rapidly.

Fill each container by gradually adding equal increments and record the depth of the water. Record the ordered pairs in a table and plot them on graph paper. Finally, connect the points with a smooth curve passing through each point.

1. Choose a container that has the shape of a cylinder. For example, use a cylindrical-shaped coffee cup, an empty juice can, or a vegetable can. When you connect the points you have plotted, what is the resulting graph? The graph is approximately a straight line, since the change is uniform.

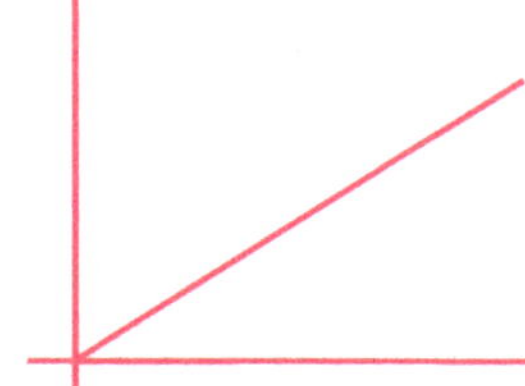

2. Choose a container that has a cone shape with the large end as the base. This could be a vase, a pitcher, or an Erlenmeyer flask. When you connect the points you have plotted, what is the resulting graph? The graph curves slowly upward at first and gets steeper at the end.

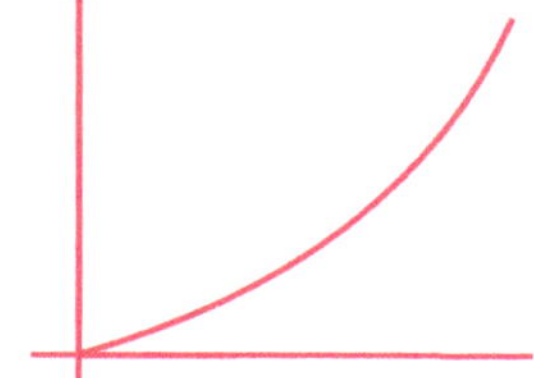

3. Choose a container that has a basic spherical shape. (See the example on p. 171.) This might be a clay pot, a drink pitcher, or a flower vase. When you connect the points you have plotted, what is the resulting graph? The graph will be somewhat like the example, depending on the container.

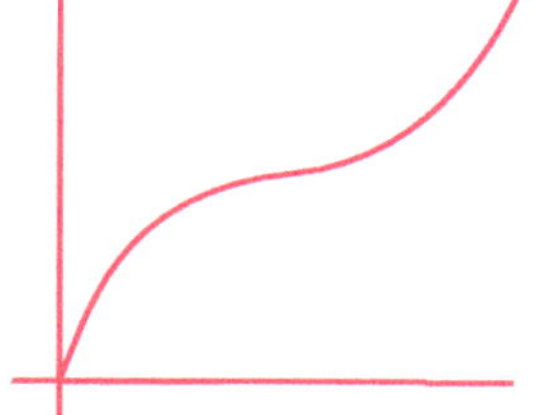

4. Choose a container that has a basic conical shape with the small end down. This will be more difficult to find, but some vases and water glasses have this shape. It is OK if it is not perfectly conical so long as the bottom starts out small and gradually gets larger toward the top. When you connect the plotted points, what is the resulting graph? The graph grows quickly at first and then increases less and less toward the end.

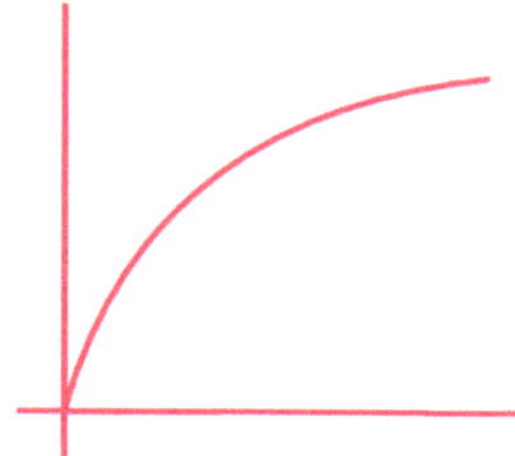

Name ________________________________

Based upon what you have learned from the graphs above, try to describe or sketch a graph that will show how the ordered pairs will plot for each of the following situations.

5. the depth of fuel oil in a horizontal cylindrical tank as a function of the amount of fuel in the tank   basically a straight line because of uniform change

6. the depth of water in a swimming pool with a uniform depth at the shallow end and a right triangle cross section for the deep end   constant (horizontal) line and then gradual increase until it reaches the maximum depth

7. the temperature of a forgotten cup of hot coffee as a function of time   rapid initial decline and then it levels off at room temperature

8. the average temperatures of New York City from January through December   gradual increase until it reaches its maximum point in July or August; then a gradual decrease each month

Using Graphs to Predict

Name _______________________________________

1. The Workout Express fitness center's membership plan requires a $200 initial fee plus $40 per month. Fitness Freedom advertises a membership plan that requires a $400 initial fee plus $30 per month. Write function rules for these membership costs. Graph the two functions on the same graph. $w(x) = 200 + 40x$; $f(x) = 400 + 30x$

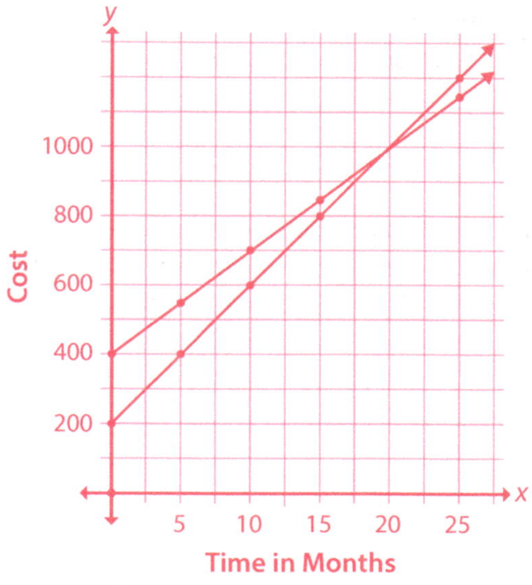

2. Find the first year's accumulated cost of membership for each fitness center in exercise 1. After how many months will the 2 plans have accumulated the same cost to be a member? What will that cost be? Which plan is best for someone who is going to use the fitness center for at least 2 years? $w(12) = \$680$; $f(12) = \$760$; 20 months; $1000; Fitness Freedom, since $f(24) = \$1120$ and $w(24) = \$1160$

3. A certain medicine requires an initial dosage of 300 mg and is absorbed in the body at 50 mg per hour after the initial dose. Write a function rule to describe the absorption of the medicine; then graph the function. $A(x) = 300 - 50x$

4. How much of the medicine in exercise 3 has been absorbed after 4 hours? How much is still remaining in the body after 4 hours? How long will it take for the entire dose to be absorbed?   200 mg; 100 mg; 6 hours

5. A certain new car cost $23,500 in November of 2003 and was worth $17,100 in November of 2005. If the car depreciates at a constant rate, what is the annual depreciation? Write a function rule to represent the depreciation of the car.
$3200; $d(x) = 23{,}500 - 3200x$

6. Graph the function in exercise 5 and use it to predict the value of the car after 5 years. According to your graph, when will the car be worthless? Do you think this is realistic? $7500; worthless in slightly less than 7.5 years; not realistic, so depreciation is probably not at a constant rate (not linear)

Formulas for Figurate Number Patterns

Name ______________________

Figurate numbers are numbers obtained from arrays of dots which have a geometric shape. These arrays have a pattern, forming a sequence with an ever-increasing number of dots. Each successive term of the sequence has more dots but keeps the same geometric shape. This provides a tool that is helpful in expressing the general term of the pattern. The general term, or formula, is an expression using the variable n to stand for the position number of the term in the sequence. Once the general formula is known, you can easily find the number of dots in a term that is a long way into the sequence and that would be impossible to draw. This formula is another example of how mathematics can make complicated things simple and laborious things easy.

Example: The following figurate numbers are called the square numbers, $S(n)$. Determine the general formula for this sequence; then find the 100th term.

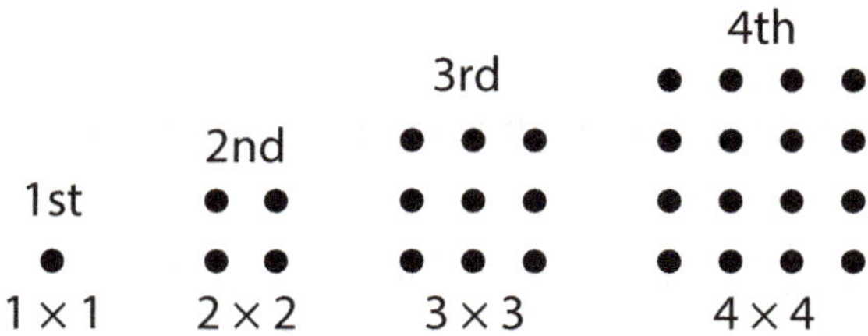

Answer:
The numbers in the sequence are 1, 4, 9, 16, . . . Notice that the numbers in the pattern match the size of the array—$1 \times 1 = 1$, $2 \times 2 = 4$, $3 \times 3 = 9$, $4 \times 4 = 16$. The nth term will be expressed as $n \times n$, so the formula for $S(n)$ is $S(n) = n \times n = n^2$. The 100th term is $S(100) = 100^2 = 10{,}000$.

It is easy to see why drawing the array and counting the dots would not be expedient. You should always check the early terms of your formula to make sure they are correct for all the terms.

Find the general formula for the following figurate numbers and answer the related questions.

1. Continue the pattern below by drawing the next set of dots.

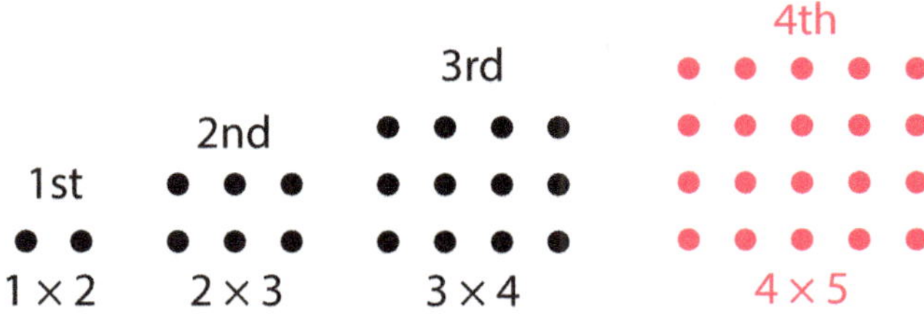

2. Write the general formula for the rectangular numbers, $R(n)$, above. $R(n) = n(n + 1)$

3. Find the number of dots in the 50th rectangular number. $R(50) = 50(50 + 1) = 2550$

4. If a rectangular array has 14,520 dots, what term is it in the sequence of rectangular numbers? (Hint: Use the problem-solving technique from Chapter 14.)
$n(n + 1) = 14{,}520$ gives $n = 120$, since $120(121) = 14{,}520$.

5. Continue the pattern below by drawing the next set of dots.

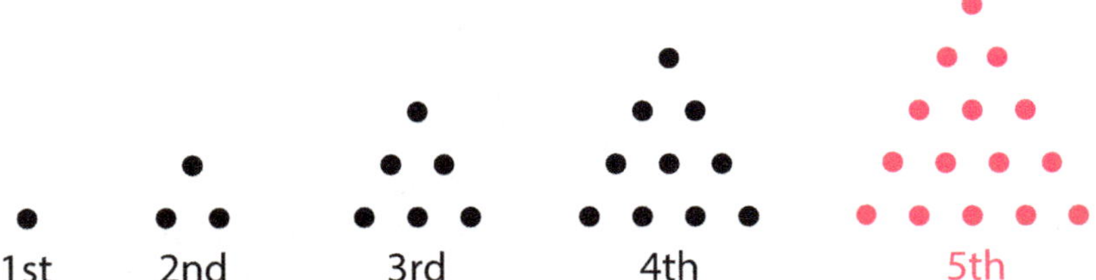

1st   2nd   3rd   4th   5th

6. Write the general formula for the triangular numbers, $T(n)$, above. (Hint: configure the dots so that they look like a right triangle instead of an equilateral triangle, and then relate them to the dots for the rectangular numbers.) $T(n) = \frac{n}{2}(n + 1)$

7. Find the 75th triangular number. $T(75) = \frac{75}{2}(75 + 1) = 2850$

8. If a triangular array has 78 dots, what term is it in the sequence of triangular numbers?
   $n = 12$

9. The pattern below represents a "train" made up of boxcars (squares) formed using matchsticks. How many matchsticks will be needed to form the diagram with 4 boxcars?   13

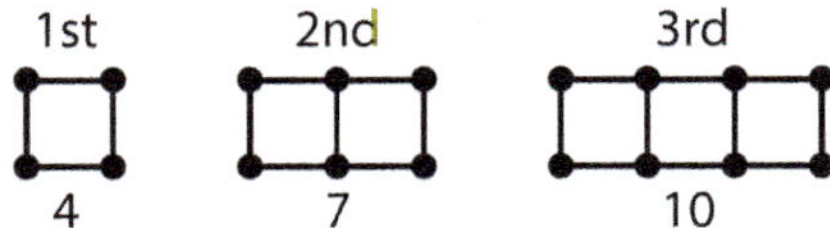

1st   2nd   3rd

4   7   10

10. Find a general formula to determine the number of matchsticks (m) it takes to form a train with n boxcars. Remember, the formula needs to work for each size train. Also, determine the number of matchsticks required to make a train with 1000 boxcars.
The difference between each term is 3, so the formula requires 3n. Adding 1 makes it work for the 1st term. Instruct the students to make sure that it works for all other terms. It does; therefore, $m = 3n + 1$. $m = 3(1000) + 1 = 3001$.

Chapter 14 Cumulative Review

Name _______________________________

D 1. List in increasing order: 36.629, 36.6099, 35.9988, 37, 36.71. **[1.1]**

    A. 37, 36.71, 36.6099, 36.629, 35.9988   **B.** 37, 36.71, 36.629, 36.6099, 35.9988

    C. 35.9988, 36.629, 36.6099, 36.71, 37   **D.** none of these

C 2. Find $2.6 + 3 \times 7^2 - 3$. **[1.7]**

    A. 51.6   **B.** 257.6

    C. 146.6   **D.** none of these

C 3. Consider the following sets. Choose the statement that is true about the sets. **[2.1]**
$A = \{1, 2, 3, 4, 5, 6\}$, $B = \{2, 4, 6, 8, 10\}$, $C = \{1, 2, 3\}$

    A. $A \subseteq C$ **B.** $1 \in A$, B, and C

    C. $\{4, 6\} \subseteq B$ **D.** $12 \in B$

A 4. Simplify $(-4)(-3 \cdot 5)$. **[2.5]**

    A. 60   **B.** -60

    C. -19 **D.** none of these

C 5. Name the property illustrated by $3.1 + (5.2 + 9) = 3.1 + (9 + 5.2)$. **[3.2]**

    A. Associative   **B.** Distributive

    C. Commutative   **D.** none of these

C 6. $(4.8 \times 10^6) \times (2 \times 10^8) = 9.6 \times 10^?$. **[3.8]**

    A. 48   **B.** 2

    C. 14   **D.** none of these

B 7. Which is the prime factorization of 384? **[4.1]**

    A. $2^2 \cdot 3 \cdot 5^2$ **B.** $2^7 \cdot 3$

    C. $2^3 \cdot 3 \cdot 4^2$ **D.** none of these

A 8. Which is the greatest common factor of 252 and 2646? **[4.2]**

    A. 126   **B.** 756

    C. 252   **D.** none of these

A 9. Subtract $29 - \frac{13}{32} - \frac{5}{8}$. **[5.2]**

    A. $27\frac{31}{32}$ **B.** $27\frac{1}{32}$

    C. $28\frac{1}{32}$ **D.** none of these

A 10. Divide $2\frac{7}{8} \div \frac{7}{8}$. **[5.5]**

    A. $3\frac{2}{7}$ **B.** $2\frac{33}{64}$

    C. 2   **D.** none of these

B 11. Translate the following sentence into an algebraic equation: "The product of 7 and a number is 28." **[6.1]**

    A. $7 + n = 28$ **B.** $7n = 28$

    C. $7 \div n = 28$ **D.** none of these

_____B_____ **12.** What range of numbers satisfies the following statement? "A number divided by 6 is less than −12." **[6.7]**

    A. $n < -2$ **B.** $n < -72$

    C. $n > -2$ **D.** $n > -72$

_____A_____ **13.** The Sharks have played 12 basketball games and won 9 of them. The Stingrays have played 15 games and won 11 of them. Which team has a higher ratio of wins to games played? **[7.3]**

    A. Sharks   **B.** Stingrays

    C. They are the same.   **D.** impossible to tell

_____C_____ **14.** If a drawing is made to a scale of 1 in. : 8 ft, what length would a line $4\frac{3}{16}$ in. represent? **[7.4]**

    A. $33\frac{7}{8}$ ft   **B.** $32\frac{3}{4}$ ft

    C. $33\frac{1}{2}$ ft   **D.** none of these

_____B_____ **15.** Find 15% commission on sales of $4050. **[8.2]**

    A. $648   **B.** $607.50

    C. $567   **D.** none of these

_____A_____ **16.** A 24 mm segment is reduced to 67% of its original size. How long is the reduction? **[8.5]**

    A. 16 mm   **B.** 8 mm

    C. 20 mm   **D.** none of these

_____C_____ **17.** Choose the most likely customary measure for selling strawberries at a store. **[9.2]**

    A. bushel   **B.** ton

    C. quart   **D.** gallon

_____C_____ **18.** If Jane is driving at 70 mph and takes her eyes off the road for 6 seconds to look at a text, how many feet will her car travel in that time? **[9.5]**

    A. 47.73 ft   **B.** 102.67 ft

    C. 616 ft   **D.** none of these

_____B_____ **19.** A parallelogram has sides of 14 and 7 units with an altitude of 5 units perpendicular to the 14 unit side. Find the area. **[10.4]**

    A. 35 sq. units   **B.** 70 sq. units

    C. 98 sq. units   **D.** none of these

_____C_____ **20.** If you know that a triangle is isosceles and one angle is 70°, which is *not* a possible measure for one of the other angles? **[10.6]**

    A. 40°   **B.** 55°

    C. 110°   **D.** All are possible.

_____A_____ **21.** Two similar figures have corresponding sides of 3 m and 10 m respectively. If the first has an area of 19 m², what is the area of the second to the nearest square meter? **[11.2]**

    A. 211 m²   **B.** 75 m²

    C. 63 m²   **D.** none of these

_____D_____**22.** A prism has right, triangular bases with sides of 5, 12, and 13 m. Find the volume if the height is 15 m. **[11.5]**

    A. 900 m³   **B.** 225 m³

    C. 487.5 m³   **D.** none of these

_____B_____**23.** If there are 2 white marbles, 8 green marbles, and 7 blue marbles in a bag, and one is drawn at random, what is the probability that it will be either green or blue? **[12.1]**

    A. $\frac{8}{17}$ **B.** $\frac{15}{17}$

    C. $\frac{7}{17}$ **D.** $\frac{13}{17}$

_____B_____**24.** If Joe has 5 shirts, 3 pairs of pants, and 2 hats, how many ways can he dress? **[12.4]**

    A. 10 ways   **B.** 30 ways

    C. 17 ways   **D.** none of these

_____B_____**25.** On a recent 10-point quiz taken by a class of 13, there were three 10s, five 9s, two 7s, two 6s, and one 4. Find the mean (nearest tenth) and the median. **[13.2]**

    A. mean: 8.1; median: 8   **B.** mean: 8.1; median: 9

    C. mean: 7.5; median: 9   **D.** none of these

_____B_____**26.** Which graph is used to show changes over a period of time? **[13.5]**

    A. bar graph   **B.** line graph

    C. histogram   **D.** none of these

_____B_____**27.** If an interval for a histogram is 75–84, what frequency would be used from the following data? 76, 83, 91, 84, 79, 86, 98, 103, 88, 79, 81, 86, 81, 83, 79, 68, 99, 74, 87, 73, 80, 78, 77, 104, 92, 79, 83, 75 **[13.6]**

    A. 16   **B.** 15

    C. 17   **D.** none of these

_____A_____**28.** Which of the following sets of ordered pairs is *not* a function? **[14.2]**

    A. (1, 2), (1, 3), (1, 4), (1, 5)   **B.** (1, 2), (2, 2), (3, 2), (4, 2)

    C. (1, 2), (3, 4), (5, 6), (7, 8)   **D.** Neither A nor B is a function.

_____B_____**29.** What is the slope of the line that joins the points (3, 4) and (−1, 6)? **[14.5]**

    A. −1   **B.** $-\frac{1}{2}$

    C. $\frac{1}{2}$ **D.** none of these

_____B_____**30.** Give the next two numbers of the sequence: 6, 8, 12, 20, 36, …**[14.6]**

    A. 52 and 84   **B.** 68 and 132

    C. 68 and 100   **D.** none of these